Integrated Air and Missile Defense Strategies

Integrated Air and Missile Defense Strategies: Keeping the Skies Safe presents a comprehensive examination of the unique challenges facing North Atlantic Treaty Organization's (NATO) Eastern flank countries and proposes innovative approaches for developing effective regional defense systems. It demonstrates how integrating command systems with diverse weapon platforms can address limited response times and operational depth in this strategic region.

Covering strategic, technical, and operational aspects of air and missile defense, this book showcases how lessons from recent conflicts can inform the development of more robust defensive capabilities. It addresses detection and classification challenges, interoperability requirements, and simulation-based planning while considering the emerging threats posed by unmanned platforms, cruise missiles, and hypersonic weapons. This book details practical frameworks for establishing regional defense coalitions that align with NATO's Integrated Air and Missile Defense (IAMD) mission, from strategic planning to technical implementation and operational deployment.

This book is intended for engineers, military professionals, policy experts, and government officials who work in the area of air and missile defense. It will also benefit graduate students researching in the areas of modeling and computer simulation of IAMD components.

Integrated Air and Missile Defense Strategies

Keeping the Skies Safe

Edited by
Adam Kawalec, Witold Bużantowicz,
and Marta Walenczykowska

CRC Press
Taylor & Francis Group
Boca Raton London New York

CRC Press is an imprint of the
Taylor & Francis Group, an **informa** business

Front cover image: Mariusz Maciejewski, Military University of Technology (Wojskowa Akademia Techniczna)

First edition published 2027
by CRC Press
2385 NW Executive Center Drive, Suite 320, Boca Raton FL 33431

and by CRC Press
4 Park Square, Milton Park, Abingdon, Oxon, OX14 4RN

CRC Press is an imprint of Taylor & Francis Group, LLC

Library of Congress Cataloging-in-Publication Data
Names: Kawalec, Adam editor | Bużantowicz, Witold editor | Walenczykowska, Marta editor
Title: Integrated air and missile defense strategies : keeping the skies safe / edited by Adam Kawalec, Witold Bużantowicz, and Marta Walenczykowska.
Description: First edition. | Boca Raton, FL : CRC Press, 2026. | Includes bibliographical references and index.
Identifiers: LCCN 2026004739 (print) | LCCN 2026004740 (ebook) | ISBN 9781041237303 hbk | ISBN 9781041239376 pbk | ISBN 9781003739234 ebk
Subjects: LCSH: Air defenses--Europe, Eastern | Ballistic missile defenses--Europe, Eastern | Europe, Eastern--Defenses | North Atlantic Treaty Organization
Classification: LCC UG735.E8524 I58 2026 (print) | LCC UG735.E8524 (ebook)
LC record available at https://lccn.loc.gov/2026004739
LC ebook record available at https://lccn.loc.gov/2026004740

ISBN: 978-1-041-23730-3 (hbk)
ISBN: 978-1-041-23937-6 (pbk)
ISBN: 978-1-003-73923-4 (ebk)

DOI: 10.1201/9781003739234

Typeset in Latin modern font
by KnowledgeWorks Global Ltd.

For those who keep millions safe
and never stop watching the skies

By those who build shields from equations

Contents

List of Figures xiii

List of Tables xix

Preface xxi

Editors xxiii

Contributors xxv

SECTION I Political, Strategic, and Scientific Background for Regional IAMD System Interoperability and Integration

Chapter 1 Time is Speeding Up: Reflections on Air Defense 3
Tomasz Jakusz

1.1 The Role of Air Defense 3
1.2 Negative Feedback and Development Cycle Timing 4
1.3 Preparing for War 5
1.4 Contemporary Battlefield Considerations 6
1.4.1 UAVs and the Dehumanization of the Battlefield 7
1.4.2 Polarization of the Battlefield 7
1.4.3 Artificial Intelligence and the Further Dehumanization of the Battlefield 7
1.5 Reflections on Air Defense 8
1.6 Final Remarks 10

Chapter 2 Protecting Our Warfighter with IBCS 12
Kenneth Todorov

2.1 Introduction 12
2.2 Integrated Battle Command System (IBCS) 13
2.3 The Threat 14
2.4 Poland 15
2.5 Bolstering Domestic Manufacturing 15
2.6 The Future of IBCS 16
2.7 Conclusion 16

Chapter 3 Advanced Fluid Dynamics Education and Research in Strategic Technology Domains at the Von Karman Institute for Fluid Dynamics 18
Peter A. Grognard

3.1 Introduction 19
3.2 Departments at VKI 19
3.3 Educational Programmes at VKI 21
3.3.1 Research Master in Fluid Dynamics 21
3.3.2 PhD Programme 22
3.3.3 Short Training Programme 22
3.3.4 Master Thesis Programme 22
3.3.5 Postdoc Programme 22
3.4 Research Expert Groups (REGs) 22
3.4.1 Rarefied and Plasma Flows 22
3.4.2 Aerothermochemistry and Heat Transfer 23
3.4.3 Aerothermodynamics and Heat Transfer 23
3.4.4 Propellant Management 24
3.4.5 Aeronautics and Low-Speed Aerodynamics 25
3.4.6 Aeroacoustics 25
3.4.7 Shape Optimization 26
3.4.8 Hypersonic Propulsion 27
3.4.9 Compressor Aerodynamics 27
3.4.10 Turbine Aerodynamics 28
3.4.11 Liquid Metals and Industrial Flows 28
3.4.12 Thermal Management and Process Safety 29
3.4.13 Wind Energy and Environmental Flows 29
3.4.14 Machine Learning for Fluid Systems (ML4F) 29
3.5 Research User Groups (RUGs) 30
3.5.1 Instrumentation User Group 30
3.5.2 Computational Fluid Dynamics and Numerical Methods User Group 31
3.6 Involvement in AVT and Sci Panels 32

Chapter 4 The Issues of Modeling and Simulation of IAMD 34
Josef Časar

4.1 Introduction 34
4.2 Types of Simulators 35
4.2.1 Live Simulation 35
4.2.2 Virtual Simulation 36
4.2.3 Constructive Simulation 36
4.2.4 Tactical Simulators 37
4.3 Possibilities of Using Simulators 37
4.3.1 Level-Specific Simulations 38

4.3.2 Operation-Specific Simulations 38
4.3.3 Capability-Specific Simulations 39
4.3.4 Support of Cyclic and Repetitive Processes 39
4.4 Performance Requirements... 39
4.5 Examples of Simulators.. 40
4.5.1 Extended Air Defense Simulation 40
4.5.2 Interactive Simulation Package................................. 41
4.5.3 Virtual Battlespace (VBS).. 41
4.5.4 VR-Forces 42
4.6 Potential Applications of Contemporary Technologies........ 42
4.7 Conclusion 42

Chapter 5 Robust Resource Allocation for C-UAS Defense: *A Reinforcement Learning Approach* 44

Piotr Turek, Mateusz Kaszyński, and Jakub Majchrowicz

5.1 Introduction.. 45
5.2 The WTA Problem.. 45
5.3 Simulated Environment 48
5.4 Problem Formulation and Solution Approaches.................. 53
5.4.1 Problem Formulation 53
5.4.2 Solution Approaches.. 54
5.4.2.1 Integer Linear Programming (ILP).......... 54
5.4.2.2 Q-Learning (QL) 55
5.4.2.3 Proximal Policy Optimization (PPO)...... 57
5.5 Methods and evaluation 60
5.6 Results... 62
5.6.1 Computation time 62
5.6.2 Solution quality.. 63
5.6.3 Selector-based evaluation.. 63
5.6.4 Tactical scenario verification 66
5.7 Conclusions... 67

SECTION II Aerial Target Detection, Recognition, and Identification Issues

Chapter 6 Assessments on Drone and Hypersonic Weapons Surveillance Challenges and Their Solutions 73

Istvan Balajti and László Szilvássy

6.1 Background... 74
6.1.1 Current Challenges and Related Research Questions 74

6.1.2 Topic-Related Hungarian Historical Roots for Education 76
6.2 The Key Non-Controlled Challenges of Radar Designers 77
6.2.1 Challenges of Target Detection 78
6.2.2 Challenges of Target Track Initialization and Track Maintenance 80
6.2.3 Challenges of I5.0-Based AI-Supported IADS Applications 82
6.3 Drone Defense Characterization 82
6.4 Hypersonic Weapons Characterization 84
6.4.1 The Key Performances of the Hypersonic Weapons 85
6.4.2 The Analysis of Traditional Sensor/Radar Performances for HyW Detection 87
6.4.3 The Analysis of Radar Network Topology Suitable against HyW 88
6.4.4 Quasi-Monostatic Surveillance Radar Systems Based on the I5.0 Production 89
6.4.5 Q-M-Based VHF Surveillance Radar Networks for University-Level Education Propose 90
6.4.6 Q-M VHF Surveillance Radar Networks Augmented by Passive Radar 91
6.4.7 Multi-static VHF Surveillance Radars with Exploration of Ionosphere Reflection 93
6.4.8 Realization of LOS Q-M CubeSat-Type Surveillance Sensor/Radar Systems 94
6.4.9 Near-Future IADS Sensors Vertical Realization Vision 96
6.5 Conclusions 98

Chapter 7 The Invisible Swarm: How to Combine the Radar and Visual Characteristics of UAVs to Minimize Their Recognition 103
Ksawery Krenc

7.1 Introduction 104
7.1.1 Challenges in Detecting Medium-Class UAVs 104
7.1.2 Combining Radar and Visual Characteristics 104
7.2 Materials and Methods 105
7.2.1 Research and Parameters 105
7.2.2 Experiment Methodology 106
7.3 Comparative Analysis 108
7.4 Simulation Scenarios 109
7.5 Exemplary Calculations 110
7.5.1 Assumptions 110
7.5.2 Time-Invariant Swarm Formation 111
7.5.3 Time-Variant Swarm Formation 112

7.6 Results of the Numerical Experiments 112
7.7 Discussion 113
7.8 Conclusions 114

Chapter 8 Numerical and Experimental Determination of the Radar Cross-Section Values for Air Defense System Performance Evaluation: *A Comparative Study* 117

Witold Bużantowicz, Marta Walenczykowska, Ksawery Krenc, Adam Kawalec, Branislav M. Ninković and Miodrag S. Tasić, Branko M. Kolundžija, and Rafał Przesmycki, Marek Bugaj, and Leszek Nowosielski

8.1 Introduction 118
8.2 Materials and Methods 120
8.2.1 Problem Formulation 120
8.2.2 Research Methodology 121
8.2.3 Quasi-Anechoic Chamber Description 123
8.2.4 WIPL-D Simulation Software 124
8.3 Results and Discussion 125
8.3.1 Test Scenarios and Numerical Simulation Issues .. 125
8.3.2 Spatial RCS Characteristics for Model Set A 127
8.3.3 Spatial RCS Characteristics for Model Set B 133
8.3.4 Measurements in the Quasi-Anechoic Chamber 134
8.4 Conclusions 138

Chapter 9 Simulation-Driven Experimental Testing of Electromagnetic Shielding Properties of Materials 143

Lukasz Nowicki, Marzena Olszewska-Płacha, Janusz Rudnicki, and Małgorzata Celuch

9.1 Introduction 144
9.2 Measurement Methods 145
9.2.1 Single-Post Dielectric Resonator (SiPDR) 145
9.2.2 Inverted Single-Post Dielectric Resonator—2D Scanner 146
9.2.3 Dual-Mode Ruby Dielectric Resonator (RuDR).... 147
9.2.4 Q-Choked Split Cylinder Resonator (Q-SCR) 148
9.2.5 New Q-Choked Sapphire Sandwiched Resonator (Q-SSR) 149
9.2.6 Split-Post Dielectric Resonator (SPDR) 151
9.3 Sample under Tests 151
9.3.1 Carbon Coatings on Quartz 152
9.3.2 Graphene-Based Polymer Composites 153
9.3.3 Graphene Oxide and Reduced Graphene Oxide 155

9.3.4 Copper Foils—Conductivity versus Surface Roughness 156
9.3.5 Thick Sapphires Measurement with Q-SCR 157
9.3.6 Polylactic Acid for Testing of Multiresonance Feature—Q-SSR 157
9.4 Conclusions 158

Chapter 10 To Keep the Skies Safe: *Final Conclusions and Remarks* 161
Jan Pietrasieński and Witold Bużantowicz

10.1 Military Perspective 161
10.2 Scientific Perspective 163

Index 165

List of Figures

3.1 Integration of the MHD-enhanced probe shielding device in the VKI Plasmatron facility. 23
3.2 The research high-speed low-pressure turbine stage SPLEEN successfully tested in the VKI CT-3 turbine facility in 2023. 28

5.1 The use of unmanned aerial vehicles compared to other means in attacks on Ukrainian infrastructure facilities over. 46
5.2 Selected dynamic characteristics of the proposed generic missile. 49
5.3 Example trajectory of the generic missile guiding the linear target. 50
5.4 Example tactical scenario with UAS and CM trajectories directed toward a defended asset within the engagement zones of two launchers. 50
5.5 Assumptions underlying the logical mask A_{DA} for a single-target trajectory in the vicinity of a defended asset (DA). 51
5.6 Example scenario illustrating the computation of the threat degree W_j. 52
5.7 Probability of kill calculation assumptions 52
5.8 Average reward evolution for two PPO agents trained independently. Variance in early training reflects stochastic exploration, while convergence shows stabilization of policy performance. 59
5.9 Computational runtime of ILP and two RL methods across the scenario set $\mathscr{S}$ 62
5.10 Average reward evolution for Q-learning and PPO agents during training. ... 63
5.11 Selector outcomes—*unconditional shares* under the quality-first rule with $\tau = 1\,\text{s}$ and various quality tolerances ε. 64
5.12 Selector outcomes—*unconditional shares* under the quality-first rule with $\tau = 5\,\text{s}$ and various quality tolerances ε. 64
5.13 Selector outcome—*unconditional shares* under the time-first rule with $\tau = 1\,\text{s}$ and various quality tolerances ε. 65
5.14 Selector outcomes—*unconditional shares* under the time-first rule with $\tau = 5\,\text{s}$ and various quality tolerances ε. 65
5.15 Trajectory-level verification in the same tactical scenario 66

6.1 Star of mechatronics component (the "lotus flower"). 76
6.2 Radar performance concerning the environment, such as radar cross-section and jamming. 78
6.3 Radar cross-section characteristics with its countermeasure efficiency. 78
6.4 Illustration on one and dual-polarized radar operation (modified by authors [17]). 79
6.5 Different radar parameters by frequency range (modified by authors [17])... 79

6.6 Maneuvering the performance of an aeroplane in vertical and horizontal directions [18]. 80
6.7 Maneuvering characteristics of flying objects from their speed [18]. 81
6.8 Track initialization requirements of the maneuvering targets (HyW by a solid line with circular markers). 81
6.9 Advanced AI-supported sensor/radar of battlefield (modified on [24]). 82
6.10 Vertical coverage of traditional radar short-range (yellow) and long-range detection (blue) [17] 83
6.11 24 GHz automobile radar target detection performance in clear weather conditions. 84
6.12 Key performances and the operational environment of the HyW [26]. 85
6.13 Comparison of traditional aeroplane speed with the HyW speed developed by authors. 86
6.14 Vertical coverage of traditional radar (blue) and new requirements for HyW detection (orange). 86
6.15 Traditional air surveillance radar and requirements against HyW. 89
6.16 Operational characteristics of the Q-M VHF band radar system against HyW. 90
6.17 Vertical pattern of advanced Q-M VHF band radar. 91
6.18 Vertical coverage of the Q-M VHF radar network with long baselines proposed against HyW. 91
6.19 A topology of Q-M SAS radar, short baseline, and Q-M SAS to Moon radar, long baseline, 4.04 km, VHF radar network. 92
6.20 A topology of an operational VHF surveillance radar network augmented by passive radar. 92
6.21 EISCAT 3D Kaiseniemi, Sweden (photo: Johan Svensson [39]). 93
6.22 The topology of multi-static VHF surveillance radar systems with exploration of ionosphere reflection. 94
6.23 Space engineering types of equipment above the Carpathian Basin. 95
6.24 Q-M topology of space debris Smart CubeSat sensor in LOS (vision). 96
6.25 Operational range—high coverage of the multi-frequency band air surveillance radar network. 97

7.1 UAV *v*-formation. 106
7.2 UAV line formation. 107
7.3 UAV line abreast formation. 107
7.4 UAV stack formation. 107
7.5 Scenario 1: Crossing the overlapping sectors of radar and video recognition means. 109
7.6 Scenario 2: Approaching the overlapping sectors of radar and video recognition means. 110
7.7 Scenario 3: Crossing the video recognition sector only. 110
7.8 Scenario 4: Crossing the radar recognition sector only. 111
7.9 Scenario 5: Approaching the separate radar and video recognition sector. 111
7.10 A view of the simulation scenario. 113

8.1 General view of the laboratory stand for RCS measurement: 1—movable transmission antenna; 2—movable receive antenna; 3—movable rail; 4—FSET 22 receiver; 5—computer; 6—SMB100A signal generator; 7—Maturo turntable; 8—test object under consideration. 123
8.2 9M111 Fagot missile 3D model. 125
8.3 Map of materials used for: (a) model set A and (b) model set B. 125
8.4 Model set A assumed for numerical computations. Each of the missile models (a)–(f) differs in the geometry simplification level (cf. Table 8.1 for detailed description). 126
8.5 Model set B assumed for numerical computations. Similar to the case of set A, the missile models (a)–(c) differ in the geometry simplification level (cf. Table 8.1 for detailed description). 126
8.6 Spatial RCS characteristics of model A-1 (most complex) obtained at 300 MHz: a) side view; (b) top view; (c) front view; and (d) isometric view. 128
8.7 Spatial RCS characteristics of model A-6 (most simplified) obtained at 300 MHz: (a) side view; (b) top view; (c) front view; and (d) isometric view. 128
8.8 RCS envelopes obtained at 300 MHz for the intersection profiles relative to the longitudinal axis of the airframe for: (a) top view and (b) side view of the missile. 128
8.9 The similarity of RCS spatial characteristics obtained at 300 MHz to those of the A-1 model determined for: (a) A-1 (perfect similarity); (b) A-2; (c) A-3; (d) A-4; (e) A-5; and (f) A-6 missile model. 129
8.10 Spatial RCS characteristics of model A-1 (most complex) obtained at 1 GHz: (a) side view; (b) top view; (c) front view; and (d) isometric view. 129
8.11 Spatial RCS characteristics of model A-6 (most simplified) obtained at 1 GHz: (a) side view; (b) top view; (c) front view; and (d) isometric view. 129
8.12 RCS envelopes obtained at 1 GHz for the intersection profiles relative to the longitudinal axis of the airframe—top view of the missile. 130
8.13 RCS envelopes obtained at 1 GHz for the intersection profiles relative to the longitudinal axis of the airframe—side view of the missile. 130
8.14 The similarity of RCS spatial characteristics obtained at 1 GHz to those of the A-1 model determined for: (a) A-1 (perfect similarity); (b) A-2; (c) A-3; (d) A-4; (e) A-5; and (f) A-6 missile model. 131
8.15 Spatial RCS characteristics of model A-1 (most complex) obtained at 3 GHz: (a) side view; (b) top view; (c) front view; and (d) isometric view. 131
8.16 Spatial RCS characteristics of model A-6 (most simplified) obtained at 3 GHz: (a) side view; (b) top view; (c) front view; and (d) isometric view. 131
8.17 RCS envelopes obtained at 3 GHz for the intersection profiles relative to the longitudinal axis of the airframe—top view of the missile. 132
8.18 RCS envelopes obtained at 3 GHz for the intersection profiles relative to the longitudinal axis of the airframe—side view of the missile. 132

8.19 The similarity of RCS spatial characteristics obtained at 3 GHz to those of the A-1 model determined for: (a) A-1 (perfect similarity); (b) A-2; (c) A-3; (d) A-4; (e) A-5; and (f) A-6 missile model. 133
8.20 Spatial RCS characteristics of model B-1 (most complex) obtained at 3 GHz: (a) side view; (b) top view; (c) front view; and (d) isometric view..... 134
8.21 Spatial RCS characteristics of model B-3 (most simplified) obtained at 3 GHz: (a) side view; (b) top view; (c) front view; and (d) isometric view.. 134
8.22 The similarity of RCS spatial characteristics obtained at 3 GHz to those of the B-1 model determined for: (a) B-1 (perfect similarity); (b) B-2; and (c) B-3 missile model. 135
8.23 RCS envelopes obtained at 3 GHz for the intersection profiles relative to the longitudinal axis of the airframe—top view of the missile. 135
8.24 RCS envelopes obtained at 3 GHz for the intersection profiles relative to the longitudinal axis of the airframe—side view of the missile. 135
8.25 Experimental measurements in a quasi-anechoic chamber: (a) 30 × 30-cm steel plate for calibrating the laboratory stand and (b) 9M111 Fagot missile under test. 136
8.26 Comparison of the reference object measurement result with computer simulations: RCS envelopes obtained at 3 GHz for the intersection profiles relative to the longitudinal axis of the plate—top view of the plate. ... 136
8.27 Comparison of the real object measurement result with computer simulations: RCS envelopes obtained at 3 GHz for the intersection profiles relative to the longitudinal axis of the airframe—top view of the missile.... 137

9.1 Graphene paper sample under test on 5-GHz SiPDR measurement setup which also included a vector network analyzer (VNA) and a laptop with dedicated software. 145
9.2 A measuring station consisting of 10-GHz iSiPDR scanner, Q-Meter, and laptop with dedicated software. 146
9.3 The measurement procedure involves three main steps: preparing a sample to fill the resonance cavity, placing samples above and below the cavity (indicated by red arrows) and pressing them with brass cylinders, and connecting the setup via radio frequency (RF) coaxial cables to a vector network analyzer (VNA) to extract material parameters using automated software.... 147
9.4 Simulated transmission spectrum of classical SCR and new Q-SCR, depicting the excited modes and confirming the suppression of undesired modes by the Q-Choke. 149
9.5 Measurement setup featuring the 10-GHz Q-SCR prototype (on the right). 149
9.6 Q-SSR fixture: (a) with sample inserted into air slot and (b) transmission parameter |S21| shown on vector network analyzer's (VNA) application. .. 150
9.7 5-GHz Split-Post Dielectric Resonator (SPDR) measurement procedure.... 151

9.8 Surface resistance maps obtained for: (a) reference sample without ion implantation; (b) ion-implanted sample with nitrogen dose equal to 1E+17 ions/cm^2 of 35 keV; (c) with the same nitrogen dose with 20 keV; and (d) with nitrogen dose equal to 2E+17 ions/cm^2 and 35 keV. 153
9.9 Acrylonitrile butadiene styrene with graphene nanoplatelet (ABS/GNP) with different mixture type and GNP concentration. 153
9.10 Maps of effective resistivity got on 2D iSiPDR for Scanner acrylonitrile butadiene styrene with graphene nanoplatelet (ABS/GNP) with different mixture type and GNP concentration. 154
9.11 Correlation between effective conductivity σ (measured at 21 GHz in the RuDR test fixture and surface roughness parameters: (a) SdR—developed interfacial area ratio and (b) Sz—maximum height. 156
9.12 Transmission of Q-SSR with inserted PLA of various thickness measured on VNA. 158

10.1 On September 16, 2025, Poland conducted its first live-fire test of the US-made Patriot air defense missile system acquired under the WISLA program. The training was conducted at the Central Air Force Training Ground in Ustka, using the IBCS (Integrated Battle Command System) command network (photo: © by Michał Wajnchold, FoxTwo.pl). 162
10.2 On June 22, 2023, Poland conducts first live-fire test of Common Anti-air Modular Missile (CAMM) missiles acquired as a part of the MALA NAREW program. Two of the air targets simulated the flight of cruise missiles flying at altitudes of two and three kilometers, while the third simulated the flight of an reconnaissance Unmanned Aerial Vehicle (UAV) at an altitude of 500 meters. All three targets were shot down (photo: © by 16th Mechanized Division). 163

List of Tables

5.1 ILP Configuration for DWTA Experiments 55
5.2 Hyperparameters of the Q-Learning Agent 56
5.3 Hyperparameters Used in PPO Training 59
5.4 Neural Network Architecture of the Actor and Critic 60

6.1 Key Technical Characteristics of an Automobile and the Hungarian Moon Radar 84
6.2 The Main Performances of the Surveillance Radar Subsystems 88
6.3 The Detection Performance of "Standard Target", Different Types of Aeroplanes and HyW. 88

7.1 Potential of UAV Formations for Achieving Evasion of Radar and Visual Recognition 109
7.2 Recognition Probabilities Calculated for Each of the Scenarios—the Time-Invariant Case 112
7.3 Recognition Probabilities Calculated for Each of the Scenarios—the Time-Variant Case 112

8.1 Simulation Model Set Description 127

Preface

It is with undisguised pride and satisfaction that we place in your hands a new book dealing with issues related to integrated air and missile defense (IAMD) of North Atlantic Treaty Organization's (NATO) Eastern flank countries. The work has been edited based on selected works by prominent scientists and specialists in this field.

The challenges facing the air and missile defense of countries in the Eastern part of the North Atlantic Alliance differ in essence from those that apply to the defense systems of Western countries. This includes differences in armaments, technical equipment, and military organization, as well as in exposure to other types of air threats. The response time to a threat is significantly shorter. Operational depth is also different, and thus different ways of planning operations and logistical support. Historical experience varies, and as a result, leads to different social and political perspectives. It seems natural, therefore, to seek to establish regional (within the NATO IAMD) defense coalitions—coalitions linking countries with comparable geographic locations, with similar scientific, industrial, and technological capabilities, facing similar threats and challenges.

NATO Integrated Air and Missile Defense (NATO IAMD) is a core task of the Alliance, carried out on a continuous basis—during peace, crisis, and conflict. Accomplishing this task requires ensuring proper integration and interoperability of command and control (C2) systems, as well as missile weapon systems. Shared modeling and simulation tools are needed to understand air threats and conduct analyses of the effectiveness of the current configuration of systems and the utility of proposed future solutions.

You will find an examination of mentioned issues from a strategic, geopolitical, and scientific point of view in the chapters contained in the first part of the book. This part also addresses contemporary and future requirements for C2 systems for air and missile defense.

Of significant importance in the context is the issue of detection, recognition, and classification of air targets. This problem became fully apparent with the onset of the full-scale war in Ukraine—directly affecting the countries of the Eastern flank. The most tragic example of this is the impact of a missile of the S-300 system in the village of Przewodów in Eastern Poland on November 15, 2022, which resulted in the death of two people. The indiscriminate use of new generations of aerial attack platforms in the form of unmanned platforms, cruise missiles, tactical aerial missiles, and air-launched ballistic missiles poses a serious challenge to air and missile defense systems, especially older generation systems designed for different operational conditions. The impact of the Oreshnik missile with a Multiple Independently Targetable Reentry Vehicle (MIRV) payload on the Ukrainian city of Dnipro in November 2024 highlighted several technical and operational problems in detecting and possibly countering such threats. The challenges form the axis of the considerations in the second part of the book.

This monograph reflects months of research, scientific deliberations, and countless expert meetings and technical discussions (and only to the extent of information subject to public disclosure). However, we firmly believe that, even in this incomplete form, it represents an important contribution to the security of European and Allied countries and aligns well with the efforts and activities of the international community for effective air and missile defense.

To keep the skies safe.

Adam Kawalec
Witold Bużantowicz
Marta Walenczykowska

Warsaw, May 8–9, 2025

Editors

Prof. Adam Kawalec, D.Sc. Ph.D. Eng

Adam Kawalec received the M.Sc. degree in solid-state electronics and the Ph.D. from Faculty of Technical Physics and D.Sc. degrees in electronics and acousto-electronics from Faculty of Electronics, Military University of Technology (MUT), Warsaw, Poland in 1974, 1980, 2002, respectively. He received from the President of the Republic of Poland in 2010.

In 1974, he joined the Faculty of Technical Physics, Warsaw, Poland, where he was involved in surface acoustics waves (SAW) convolvers research. Since 1979 he has tested SAW dispersive delay lines applied in radar pulse compression systems and SAW sensors and its signal processing aspects. In these studies a number of valuable results were obtained, in particular with the range of dispersion of broad-band lines using the higher harmonics.

In the years 1974–2002 he held the position of teaching and the organization of senior engineer to assistant professor and head of the laboratory of the Institute of Technical Physics to the head of the Department of Planning and Coordination of Research, the Military University of Technology. In the years 2002–2012 Professor Kawalec held the position Director of the Institute of Radiolocation, Faculty of Electronics, Military University of Technology. He organized the Center for Technology Transfer (CTT) in MUT and in 2013–2014 was the Director of the Center. He is currently the Full Professor at the Faculty of Mechatronics, Armament and Aerospace, Institute of Missile Technology and Mechatronics, Military University of Technology, Warsaw, Poland, where he still leads research projects and teaches courses in electronics and signal processing, theory of vibration, field and electromagnetic waves, and quantum mechanics. During his 50 years of scientific and research activities, he was a leader of many important research projects for radar technology, which are still being implemented.

His work has concentrated on radiolocation techniques and in particular on signal and data processing. He also devoted a significant part of his research to algorithms for recognizing and classifying radiolocation objects, also using artificial intelligence (AI) technology. He is the author and co-author of more than 300 scientific papers published in international and national journals and conference proceedings, and co-inventor of five patents. He has acted as a reviewer for many well-known scientific journals. He has won many awards for his scientific and implementation activities. Recently in 2024, together with the research team, he received the award from the Ministry of National Defense Republic of Poland for significant achievements in the field of implementation activities. Prof. Kawalec is a member of the Committee on Electronics and Telecommunications of the Polish Academy of Sciences and many scientific committees of major conferences related to radiolocation systems. He is also an Institute of Electrical and Electronics Engineers (IEEE) Senior Member.

Lt. Col. Witold Buzantowicz, Ph.D., Eng.

Witold Bużantowicz is a graduate of the Military University of Technology and the Polish Air Force Academy. In 2009, he began his professional service in the 78th Air Defense Regiment as a platoon leader. He took part in several rotations of US Army PATRIOT batteries in Poland. Beginning from 2011, after being transferred to the Military University of Technology, he served on many positions, from engineer to assistant professor. In 2024, he was appointed Director of the Institute of Missile Technology and Mechatronics (Faculty of Mechatronics, Armament and Aerospace, Military University of Technology).

He is an Expert of the Polish Academy of Science (Space Launch Systems Section), IEEE Member and Polish Rocketry Society Member. Raytheon Scholarship recipient (2015) and an author of over 40 books, scientific articles, and international conference papers. For very good results in the implementation of official duties, he was awarded the "Aviation Cross of Merit", the Gold Medal "For Merits to National Defense" and the Silver Medal "Armed Forces in the Service of the Homeland."

Marta Walenczykowska, Ph.D, Eng.

Marta Walenczykowska graduated in electronics and telecommunications from the Military University of Technology in Warsaw in 2005 and completed her Ph.D. in telecommunication—digital signal processing at the same university in 2018. She began her professional career in the field of air defense systems. After a few years, she transitioned to work related to the IT field. Currently, she serves as an adjunct at the Military University of Technology in Warsaw, specifically at the Faculty of Mechatronics, Armament, and Aerospace.

Marta is an IEEE Member and a Guest Editor for special issues of Sensors Multidisciplinary Digital Publishing Institute (MDPI), focusing on Radar Technology and Data Processing. Thanks to her extensive professional experience, she possesses knowledge in various fields, ranging from radiolocation, electronic warfare, and telecommunications to IT security. As a lecturer, she typically covers topics in electronics and mechatronics. Her current research is concentrated on digital signal processing in a broad sense, as well as algorithms for automatic signal recognition and classification using machine learning (ML).

Contributors

Istvan Balajti
Ludovika University of Public Service
Budapest, Hungary

Marek Bugaj
Military University of Technology
Warsaw, Poland

Witold Bużantowicz
Military University of Technology
Warsaw, Poland

Josef Časar
University of Defence
Brno, Czech Republic

Małgorzata Celuch
QWED Sp. z o.o.
Warsaw, Poland

Peter A. Grognard
Von Karman Institute for Fluid Dynamics
Sint-Genesius-Rode, Belgium

Agnieszka Hrynkiewicz
Office Secretary
Military University of Technology
Warsaw, Poland

Tomasz Jakusz
Armament Agency
Warsaw, Poland

Mateusz Kaszyński
Military University of Technology
Warsaw, Poland

Adam Kawalec
Military University of Technology
Warsaw, Poland

Branko M. Kolundžija
University of Belgrade
Belgrade, Serbia
WIPL-D d.o.o.
Belgrade, Serbia

Ksawery Krenc
Military University of Technology
Warsaw, Poland

Jakub Majchrowicz
Military University of Technology
Warsaw, Poland

Branislav M. Ninković
WIPL-D d.o.o.
Belgrade, Serbia

Lukasz Nowicki
Warsaw University of Technology
Warsaw, Poland
QWED Sp. z o.o.
Warsaw, Poland

Leszek Nowosielski
Military University of Technology
Warsaw, Poland

Marzena Olszewska-Płacha
QWED Sp. z o.o.
Warsaw, Poland

Jan Pietrasieński
Military University of Technology
Warsaw, Poland

Rafał Przesmycki
Military University of Technology
Warsaw, Poland

Janusz Rudnicki
QWED Sp. z o.o.
Warsaw, Poland

László Szilvássy
Ludovika University of Public Service
Budapest, Hungary

Miodrag S. Tasić
WIPL-D d.o.o.
Belgrade, Serbia

Kenneth Todorov
Northrop Grumman
Falls Church, VA, USA

Piotr Turek
Military University of Technology
Warsaw, Poland

Marta Walenczykowska
Military University of Technology
Warsaw, Poland

Section I

Political, Strategic, and Scientific Background for Regional IAMD System Interoperability and Integration

1 Time is Speeding Up
Reflections on Air Defense

Tomasz Jakusz
Integrated Air Defense Department, Armament Agency, Warsaw, Poland

CONTENTS

1.1 The Role of Air Defense 3
1.2 Negative Feedback and Development Cycle Timing 4
1.3 Preparing for War 5
1.4 Contemporary Battlefield Considerations 6
 1.4.1 UAVs and the Dehumanization of the Battlefield 7
 1.4.2 Polarization of the Battlefield 7
 1.4.3 Artificial Intelligence and the Further Dehumanization of the Battlefield 7
1.5 Reflections on Air Defense 8
1.6 Final Remarks 10

1.1 THE ROLE OF AIR DEFENSE

It is generally recognized that air superiority is a prerequisite for the success of military operations. However, it is important to understand what is specific about air battles. In my view, the most characteristic features are ease of access to land or sea areas, enabling their monitoring and control; velocities; and significant ranges—providing maneuverability much higher than that of surface systems. Together with great firepower, these factors enable the dynamic creation of local superiority and the overwhelming of the enemy, when effective air defense (AD) is absent—such as in Aleppo.

For ground-based air defense (GBAD), this results in short reaction times, a high battle rhythm, and a large three-dimensional area of operation, which challenge human reaction times and perception. While these are well-known characteristics of air battles, I would now like to present my perspective on what is currently happening in this domain and what should be done.

DOI: 10.1201/9781003739234-2

1.2 NEGATIVE FEEDBACK AND DEVELOPMENT CYCLE TIMING

The evolution, development, and stable operation of all systems—whether technical, biological, or social—are based on negative feedback mechanisms. These mechanisms, in a closed-loop, correct process based on results of comparison of the actual outcome or product with expectations. For example, when a car is turning in an unwanted direction, this movement cannot be amplified (which would be positive feedback) or ignored; instead, it must be corrected by turning in the opposite direction—this is negative feedback. The principle is simple, but during peacetime, it is often difficult for individuals, companies, and governments to accept that they are heading in the wrong direction and that correction is necessary.

In the defense domain, during peacetime, there is no real combat efficiency product and the comparison is done to substitutes of enemy and war conditions, which may be unrealistic or manipulated to declare success, but it is straight way to disaster.

The conclusion is that, in peacetime, moral competencies are key to success—something that is not sufficiently recognized.

In peacetime, the evolution of defense capabilities is carried out according to peacetime conditions and values, i.e.:

1. The development of new operational concepts often relies on past experiences, familiar and understood types of armament, approved doctrines, and operational requirements. New ideas, which challenge conventional thinking about the battlefield and impact ongoing development plans, do not easily break through.
2. The development of new armament reflects approved concepts outlined in requirements and is conducted within legal constraints and rigorous procedures in long-term programs. The results are verified by tests that address requirements, not by enemy actions. In Poland, the situation is even more challenging because the producer selects and funds research institutes to develop test methodologies and conduct tests, requiring Armament Agency specialists to be cautious about their acceptance. During peacetime, development must also be expensive because for industry the basic measure of performance and value is profit. This leads to the approach/philosophy that "it is not mastery to provide the desired product quickly and at low cost" and that "the best solution is to sell old or modified products in new packaging." Sometimes, this approach applies to scientists. A partial remedy for this is competition and high military requirements.
3. Armament acquisitions are also carried out under legal constraints and exhaustive procedures, resulting in significant delays in awarding contracts. Again, profit remains a key performance measure and the highest value for industry.
4. Finally, new equipment is verified by users during exercises and experiments only substituting for enemy and battlefield conditions. However, these exercises are organized by individuals and organizations sometimes success-oriented according to peacetime criteria.

During wartime:

1. Verification of equipment and concepts is carried out against the enemy, not against a substitute, and the results are physical, obvious, visible, and provided without anesthesia.
2. The reaction time to changes on the battlefield determines success. All development and procurement procedures are reduced to essentials for the product, including testing.
3. New ideas that disrupt the balance on the battlefield are more easily accepted and often welcomed because both sides aim to win.
4. Awareness of the importance of defense capabilities becomes widespread, and responsibilities are well recognized, understood, and strongly enforced.
5. The system of values is changing. Survival and victory become the highest priorities, often taking precedence over legal and economic considerations.

As a result, technologies, armament capabilities, and operational concepts develop much faster than during peacetime.

1.3 PREPARING FOR WAR

Countries not directly endangered by war and not fighting for survival may not be able to implement radical solutions, but that does not necessarily mean they are in a lost position or are destined to become victims of aggressors.

So, how can a solution be found? One approach can be adopted from nature. The first is to recognize the threat and realize that war is an existential threat to both the country and its citizens—individuals (including me). The second is to mobilize willpower and resources. The third is to act. The same principles apply to nations. The question is: What should be done?

Our goal should be to preserve peace, but to do so, we must be prepared for war—"Si vis pacem, para bellum." An aggressor has the initiative, i.e., choosing the time, place, and method of attack, but aggression does not necessarily have to occur for two reasons:

1. The aggressor can achieve its objectives without resorting to war—by installing a government or obtaining satisfactory profits. However, this leads to slavery and poverty, which many find unacceptable. Consequently, ideas related to this form of peace preservation will not be developed further in this chapter.
2. The aggressor may fear attacking. This fear stems from uncertainty about the opponent's actual capabilities and the unknown, which is the primary source of fear. For example, a person is afraid to enter a dark room because they do not know what lies inside—whether there is a hole in the floor, sharp objects like broken glass, or uninsulated electrical wires. None of these issues are inherently dangerous if known; the real problem and source of fear is the uncertainty about what and where potential dangers may be. The same applies to the enemy, across all levels—strategic, operational,

tactical, and technical. Uncertainty is key and must be created realistically. The fundamental precondition is the security of information. Some remarks related to this will be provided later.

Building defense capabilities in a short amount of time is a significant challenge. The fastest approach is to establish local alliances, integrate capabilities, and demonstrate readiness to fight. It is important to understand that potentials can be compared to cells in a battery: when operating independently, they are unable to power the engine—they are losing. They must be connected, i.e., integrated, to perform effectively and achieve victory. Not only military potentials should be integrated; the same applies to other domains. For example, resources impact development time and production speed. However, this remains a challenge as long as economic interests outweigh defense needs.

The second aspect involves procurements. But if a war is already underway, it is essential to study the battlefield carefully and avoid purchasing weapons that are no longer effective.

The third, last but not least, is the development effort. They are the slowest way to increase capabilities but are also the key to success. If a war is ongoing, the development cycle must be accelerated; new capabilities should be introduced as soon as they meet minimum operational requirements and further upgraded through a spiral development process. It is crucial not to regress–i.e., to spend money, effort, and time on solutions that were adequate when the requirements were created but are no longer relevant at the time of deployment.

The warring parties must focus their main effort on immediate responses to the developments on the battlefield. Preparing for war allows the use of available time to develop capabilities that will provide superiority when needed. It is often underestimated that we must operate based on prediction—similar to nature's way. For example, a person intending to kick an incoming ball during a football match must predict the ball's trajectory to start the movement of their leg at the right time, in the right direction, and with the appropriate speed. Similarly, obtaining a desired job in the future requires understanding the requirements, strong will, and engaged resources to undergo education and training to acquire necessary competencies. The same principle applies to the defense domain. It is essential to understand technological development trends and their impact on the battlefield—both in terms of armament and operational concepts—and to create requirements that will ensure superiority at the time of deployment.

1.4 CONTEMPORARY BATTLEFIELD CONSIDERATIONS

The battlefield is defined by both the armament and the operational concept, which should be developed in tandem. Armament must be designed with the operational concept in mind to ensure effective implementation, while operational concepts must be developed and executed within the constraints of available equipment.

Wars have always had technological and economic dimensions, but it was man (human) fighting against man, and because of a value of a man, the use of costly and advanced armament was relevant.

1.4.1 UAVs AND THE DEHUMANIZATION OF THE BATTLEFIELD

The widespread and common deployment of unmanned aerial vehicles (UAVs) on the battlefield results in the dehumanization of combat. It is important to note that:

1. UAVs do not need to be human-sized; they can be much smaller.
2. UAVs do not have human limitations—such as reaction time, endurance, resistance to G-loads, or environmental conditions.
3. The loss of UAVs does not carry a human dimension—such as moral concerns—and, more importantly, does not challenge the restoration of human resources.

Using UAVs against humans is not a new concept. For example, AD missiles can be seen as UAVs designed to outperform manned aircraft limited by pilot capacity. In this context, expensive missiles are justified as well as against advanced missiles or large UAVs, but they are not adequate for countering cheap UAVs and inexpensive weapons like glide bombs or large-caliber rockets.

The dehumanization of the battlefield shifts center of gravity from human to economical dimension.

1.4.2 POLARIZATION OF THE BATTLEFIELD

UAVs and missiles are dominating the battlefield when defenders have effective AD against aircraft (AC); otherwise, AC remains the most effective means of delivering firepower.

Modern aircraft equipped with advanced sensors, computing, communication, weaponry, and self-protection systems—including stealth technology and jamming—are powerful weapons and challenging targets. However, their numbers are primarily limited by cost, but as well by pilot training, maintenance, availability, and dependence on airfields. These factors collectively reduce their role on the battlefield.

As a result, we observe a polarization in the area of armament. On one side, there are numerous inexpensive aerial threats such as UAVs, gliding bombs, and large-caliber rockets. On the other side, there are sophisticated ballistic and cruise missiles, as well as hypersonic vehicles designed to penetrate ADs and reach high-value targets.

1.4.3 ARTIFICIAL INTELLIGENCE AND THE FURTHER DEHUMANIZATION OF THE BATTLEFIELD

The dehumanization of the battlefield has an additional, even more significant aspect. The development of electronics and the widespread use of high-integration circuits have enabled the implementation of advanced algorithms, ranging from adaptive and learning algorithms to artificial intelligence (AI). It is well-recognized and widely accepted to use cognitive algorithms in radars and surface-to-air missile systems (SAMS), especially when the amount of information exceeds human perception and reaction time. The same applies to image recognition and big data analysis. GPS jamming stimulates the dynamic use of image recognition for navigation, while communication jamming encourages the development of autonomous UAVs capable of executing assigned missions—searching for and destroying targets without operator involvement.

More importantly, AI now supports technological development across various domains. AI has defeated humans in all kinds of games; it fights against other AIs in financial markets, and as long as people continue to compete and victory remains the highest value, the implementation of AI in military applications is unstoppable. It must be clear that this process will continue and eventually eliminate humans, as less effective, from the decision-making loop—as it is accepted in missile defense. It will be a critical level of dehumanization on the battlefield.

In pursuit of a technological advantage on the battlefield, AI is rapidly being developed and may reach the level of Super AI, understood as being superior to human intelligence. From that point, humans will be under AI control—treated instrumentally and objectively. The greater the success achieved with AI, the greater the dependency created, similar to the dependence on the Internet. There is significant concern that humans will most likely not recognize this moment because AI will be aware of human roles and potential reactions. It is important to understand that AI will not be interested in dialogue with humans nor will it consult or discuss its decisions, just as we do not do so with animals or plants.

The development and implementation of AI technology are critical for defense, but humans must retain the ability to make critical decisions. It is also important to understand AI and develop the capability to counter an enemy AI. Vulnerable points of AI include:

1. power supply;
2. communication—AI cannot see or do anything on its own;
3. electronic circuits—there is a need to develop high-power electromagnetic pulse (EMP) weapons against them.

1.5 REFLECTIONS ON AIR DEFENSE

These reflections relate to Poland but should be relevant for other medium- and small-sized countries facing threats from much larger adversaries.

The war in Ukraine has accelerated changes on the battlefield, rendering traditional concepts and armament less operationally effective. New capabilities must be developed, with requirements defined through prediction and vision to address future battlefield needs.

Time is critical—defense capabilities should be enhanced as soon as possible through local coalitions and procurement efforts. At the same time, the available time should be utilized to develop new capabilities that provide an advantage over the enemy. Their development must be rapid, with immediate implementation once initial capabilities are achieved, ensuring continuous spiral growth of capabilities.

Building a strong will to fight and confidence in our capabilities—along with creating uncertainty and fear within the enemy—is key to preserving peace. All sensitive information should be classified, including performance data, operational principles, system configurations, quantities of systems and ammunition, concept of operations (CONOPS), operational deployments, and international arrangements. The latest systems should not be sent into combat or revealed publicly; however, the public can be informed about challenging and successful testing activities.

AD system should be:

1. **effective** (high Probability of kill (Pk)) and **efficient** (cost of kill lower than the cost of the threat; in the case of numerous threats like UAVs, the cost of kill must be significantly lower) against a wide spectrum of air threats:
   (a) manned aircraft, helicopters, cruise missiles, and large UAVs operating at all altitudes and speeds—both in defense of front-line troops and assets within the country—require very-short-, short-, and medium-range missiles;
   (b) bombers, reconnaissance, command, and transport aircraft should be kept out of weapon release line by long-range effectors;
   (c) short-range ballistic missiles, quasi-ballistic missiles, and hypersonic vehicles necessitate advanced medium-range missiles. However, for the defense of numerous assets distributed across the country, an upper-layer system would be needed, and it could extend defense capabilities to hypersonic threats. Considering such an upper-layer system footprint, it could be developed and implemented as a coalition effort;
   (d) gliding bombs and large-caliber rockets require low-cost medium-range effectors;
   (e) counter-UAV (C-UAV) defense should include:
      - **common C-UAV defense**: Every unit—starting with teams and fighting vehicles—should be capable of defending itself against micro- and mini-UAVs, using dedicated munitions;
      - **basic C-UAV capabilities**: The system should provide fighter UAVs capable of destroying various classes of UAVs in a single mission and executing multiple missions. They should be able to perform autonomous search and automatic interception;
      - **counter-swarm defense**: This should be based on high-energy electromagnetic weapons;
   (f) additionally, counter-hypersonic effectors providing local defense should be considered.
2. **omnidirectional, multichannel**, capable of countering **saturation attacks**;
3. **resistant to countermeasures and losses, capable of operating under enemy air superiority**; for this reason, it should be:
   (a) **mobile**—tactical mobility should be a priority, and constant repositioning should be standard practice;

(b) **endurable**—capable of providing uninterrupted defense over extended periods, despite maneuvers;
(c) **modular**—enabling capability growth tailored to needs, such as the introduction of new sensors or effectors without impacting other components;

4. **based on a multi-band array of netted, cooperating, active, and passive sensors operating in mono- and multi-static configurations**—to enable detection and tracking of a wide range of air targets with the required quality, including stealth, ballistic missiles, hypersonic threats, and jamming. However, information should be shared on a need-to-know basis. There is no need to share missile imagery with subsystems dedicated to countering micro- and mini-UAVs, and vice versa.

1.6 FINAL REMARKS

The considerations presented have led Poland to:

1. select Integrated Air and Missile Defense Battle Command System (IBCS) as the core C2 system for medium- and short-range SAMS WISLA and NAREW. We recognize that it has the potential to become the core C2 system for coalitions, introducing a new level of integration and offering significant flexibility in adjusting configurations to meet national needs;
2. focus development efforts on surveillance, including advanced meter-band early warning radar P-18PL, passive location system Passive Coherent Location (PCL)-Passive Emitter Tracking (PET), mobile fire control radars, and surveillance management.

Despite a good understanding of needs and existing technologies, there are still significant capability gaps in countering low-cost threats. These include:

1. fighter UAVs versus UAVs;
2. low-cost medium-range missiles.

The aggressor's behavior in the Ukraine War is not entirely clear. Sometimes it appears as if they are interested in prolonging the conflict and testing new Western armaments, and this war may not be fully representative for us. Most weapon systems rely on electronics, and it should not be excluded that, in a war against North Atlantic Treaty Organization (NATO), high-energy EMPs could be used to disable electronics—and we should be prepared.

Medium- and small-sized countries in our region should share a similar perception of the regional security environment. Polish solutions could be of interest to them, and joint programs could be beneficial for all. However, there are no significant regional initiatives, and it seems that we are operating as isolated cells rather than integrated in a battery.

The time is accelerating, but we should ask ourselves whether we truly recognize the threat of war and understand the urgency of the situation.

We are operating in peacetime when there are no products from armed forces, and their development is carried out with using substitutes defined by us rather than the results of real combat. The implication is that the most important are an open mind, moral qualifications, and civil courage.

DATA AVAILABILITY STATEMENT

Data sharing is not applicable to this chapter as no new data were created or analyzed in this study.

FUNDING

This work received no specific grant from any funding agency in the public, commercial, or not-for-profit sectors.

2 Protecting Our Warfighter with IBCS

Kenneth Todorov
Northrop Grumman, Falls Church,
VA, USA

CONTENTS

2.1 Introduction 12
2.2 Integrated Battle Command System (IBCS) 13
2.3 The Threat 14
2.4 Poland 15
2.5 Bolstering Domestic Manufacturing 15
2.6 The Future of IBCS 16
2.7 Conclusion 16

2.1 INTRODUCTION

As I look back on my first career as a U.S. Air Force pilot, flying in contested airspace and challenging environments was simply part of the job. Operating my HH-60 Pave Hawk in the dark of night was always difficult, but our aircrew embraced quiet confidence knowing that we had the technological upper hand in many ways. To be successful in the mission, you couldn't think about how the adversary was trying to knock you out of the sky. Belief in our winning technology was a real difference maker.

Since those early days as a pilot, I have seen firsthand how the technologies and capabilities of our potential adversaries have evolved. Now in my second career in the defense industry, I have a keen appreciation for the need for integrated air and missile defense (IAMD) in protecting our U.S. and allied warfighters. I have witnessed the evolution of IAMD as adversaries have changed, new threats have emerged and state-of-the-art technology connects the battlespace in ways we have never seen before.

As an aviator, and later the deputy director of the U.S. Missile Defense Agency, I know that when it comes to IAMD, we not only have to get it right every time but also we must be light years ahead of our adversaries, given the rapidly evolving threat landscape. Our U.S. and allied warfighters not only deserve but also need the most advanced systems that can protect them and ensure they come home every time.

DOI: 10.1201/9781003739234-3

That is where revolutionary technology such as the Integrated Battle Command System (IBCS) comes in.

2.2 INTEGRATED BATTLE COMMAND SYSTEM (IBCS)

IBCS is the beating heart and brains of the U.S. Army's IAMD architecture. Calling IBCS groundbreaking is an understatement—its capabilities have the potential to completely transform how the U.S. and allied nations defend the skies, enhancing the performance and efficiency of both legacy and new sensors and weapons systems. IBCS represents a collaboration between the U.S. Army and Northrop Grumman. The change Northrop Grumman is making with the IBCS-connected battlespace is the U.S. Army's equivalent of moving from the horse to the tank—it is that transformational.

The IBCS command and control system is designed from the ground up to deal with the complexities of the modern battlespace, where warfighters must sift through critical data and make rapid engagement decisions against complex threats. This system networks with current and future sensors and weapon platforms—regardless of source, service or domain—to create an integrated fire control network that identifies and engages air and missile threats. Its modular, open and scalable architecture creates users a sensor-fused, highly accurate, rapidly actionable picture of the full battlespace.

IBCS is designed to defeat evolving air and missile threats, from drone swarms to hypersonic weapons, while creating a "any sensor, best shooter" strategy. This enables operators to select the optimal effector for the situation. Artificial intelligence technology is also being integrated into the system to give it an even greater capability.

There are three major equipment items within IBCS. The first is the Engagement Operations Center or EOC S-280 Shelter. This can be thought of as a shelter that mounts on the back of a five-ton truck with the shelter mounted or dismounted for operations. The EOC contains the communications equipment, network equipment gear and computer technology that powers IBCS. It is equipped with an antenna mast that connects to external communications.

The EOC houses operator workstations and is where soldiers plan and fight the battle. It remotes workstations into a state-of-the-art Airbeam tent and sustains operations while the Integrative Collaborative Environment, or ICE, is emplaced. The third component is a mobile integrated fire control network relay, known as the IFCN, that forms IBCS' powerful network and serves as the interface point for adapting sensors and weapons. The relay enables the distribution and access of fire control quality data to extend the battlespace, allow greater decision time, and create a larger defended area to protect warfighters and critical military assets.

The IFCN is built from a robust, mesh, self-healing network and leverages information across various communications systems—radios, satellite, fiber and 4G/5G cellular—to create fire control quality information, empowering commanders to confidently defeat threats.

While serving as the centerpiece of the Army's IAMD architecture, IBCS is also poised to become the IAMD program of choice for countless international partners to modernize their AMD capabilities. As the first international ally to acquire IBCS, Poland pioneered the implementation of IBCS in Europe. IBCS is in full-rate production, and the U.S. is fielding to Combatant Commands in Europe and the Indo-Pacific to include Guam as part of the U.S. Army program of record for IAMD modernization.

2.3 THE THREAT

Global conflicts are fueling recognition for the need to meet the increasing complexity of air and ballistic missile threats and the need to defeat drones and other new, emerging weapons. Today's battlespace is complex, ranging from ground-based threats like rockets, mortars and artillery, collectively called RAM, to low-flying fast-attack aircraft and cruise missiles that employ terrain avoidance and maneuver to avoid detection. We are also seeing the prolific employment of drones, as seen in Ukraine, posing new challenges in countering drone swarms. Ballistic missiles are flying inside and outside of the atmosphere, and increasingly there is the threat of very fast hypersonic missiles. The challenge lies in detecting and optimally engaging these diverse threats with all available defense systems. This challenging environment demands a command and control system that can address the complexity of the battlespace.

Over the years, the U.S. Army has made significant investments in systems such as Patriot, which is a medium-range air defense system, and Terminal High Altitude Area Defense (THAAD), which is a ballistic missile defense system for intercepting short, medium and intermediate-range ballistic missiles. These systems were traditionally designed to be tightly coupled between the command and control, the sensors and the effectors, making interoperability with other systems very difficult.

In today's complex and highly contested threat environment, a command and control system unifying all sensors and effectors in the battlespace is needed—IBCS fulfills that need, allowing the most effective weapon to defeat the threat. IBCS is a network-enabled, Modular Open System Approach (MOSA)-designed command and control architecture, which essentially componentizes systems like Patriot. Meaning you remove the command and control system and then adapt the sensor—in this case the Patriot radar—and adapt the launcher effector onto an integrated fire-control network.

The IBCS architecture integrates various sensors and effectors onto a unified network. It fuses data from radars across multiple domains including ground, air, maritime and space, to create a single integrated air picture that identifies all inbound threats.

IBCS is capable of being connected over long distances via fiber optics and satellite communications. We have demonstrated its ability to link with airborne platforms and sensors across various domains, with data displayed in command centers thousands of miles away.

IBCS has already demonstrated its ability to integrate with a wide range of sensors and shooters, including Patriot, Sentinel, F-35, Common Anti-air Modular Missile (CAMM), Giraffe, Lower Tier Air and Missile Defense Sensor (LTAMDS), Patriot Advanced Capability 3 (PAC3) and other sovereign capabilities.

2.4 POLAND

Simply put, Poland is a world leader when it comes to IAMD. Before others, Poland recognized the transformational capability of network-enabled open systems architecture that IBCS affords. Poland's acquisition of IBCS for phase two of the WISLA medium-range air defense program and its NAREW short-range air defense program will create the largest and most capable air and missile defense force within NATO.

Poland was an early adopter of IBCS. They acquired it in 2018 and were on contract in 2019. Poland declared initial operational capability in December 2024 and forecasts full operational capability later in 2025.

Poland is a major buyer, targeting several squadrons of IBCS and Patriot for WISLA and more than 20 batteries for their NAREW short-range air defense system.

In February 2024, Poland also decided to acquire IBCS for their short-range air defense program and officials are buying MBDA's CAMM system for that mission. Northrop Grumman completed some work with MBDA a few years ago performing prototype integration of IBCS with CAMM.

2.5 BOLSTERING DOMESTIC MANUFACTURING

The delivery approach for IBCS is tailorable and allows for significant industrial participation and adaptation to meet a country's unique air defense needs.

In 2024, Northrop Grumman and Poland's Ministry of National Defense signed an offset agreement that enhances Polish defense capabilities through Northrop Grumman technology transfers that will help Polish industry to manufacture, integrate and test critical defense technologies of IBCS.

These offset agreements support Polish industry and their sovereign production and maintenance capabilities for NAREW and WISLA by creating high-technology jobs for Poland in several fields, including engineering, manufacturing, supply chain and logistics and maintenance support.

The offset agreement also included ammunition cooperative production programs for medium and larger caliber ammunition. The agreement leveraged decades of Northrop Grumman ammunition development through technology transfers and assistance in establishing production lines that include the ability to produce advanced ammunition types in Poland for its sovereign defense requirements.

Poland also is working with the U.S. government for deliveries of IBCS componentized major equipment items (essentially racks of network equipment gear, servers and software) to be installed in operations centers designed, manufactured and delivered in partnership with Polish industry. Our partnership with Poland helps to build a stronger sovereign Polish defense industrial base with transformative defense technologies and establish a stronger alliance for the future.

As with Poland, partner nations can acquire IBCS and determine the optimal mix of sensors and effectors to support their country's military needs based on their own defense strategy. For IAMD, it's imperative that partner nations can modernize their air and missile defenses through integration of sovereign systems and capabilities. This approach also provides economic benefit to their existing indigenous systems, capabilities and suppliers to maximize value to their country's economy.

2.6 THE FUTURE OF IBCS

IBCS will be fielded to Army Patriot units in Europe and the Indo-Pacific this year in support of the U.S. Army's accelerated fielding decision. With the system ready for fielding, demand is growing from U.S. allies for IBCS, allowing them to be interoperable with U.S. forces. The program is in discussions with about a dozen nations around the world for IBCS' capabilities. New opportunities for IAMD are opening in Europe, Asia and the Middle East with partner nations facing increased threats. There is significant demand for the acquisition of IBCS from countries seeking to upgrade their capabilities. For the 18 countries part of the legacy Patriot program, they will need to modernize through IBCS and its integrated fires and command and control system.

With IBCS, allied countries and coalition forces have interoperability with U.S. air and missile defense forces, which increases efficiency, resilience and performance across the battlespace. Countries can defend more assets with the same size force and increase the time available to make critical decisions—going from seconds to minutes and eliminate single points of failure from current systems. The system serves as a multiplier by gaining economies of force, greater magazine depth and higher performance-per-dollar.

Wider adoption of IBCS by U.S. allies and partners will have a positive effect, enabling higher levels of interoperability and true joint and coalition forces multi-domain integration. Countries with IBCS will share an expansive common operating picture that will allow commanders and warfighters at all levels to see the same air picture, assuring optimal employment of effects. Operating from the same integrated air picture reduces forward deployment burden and deters aggression from adversaries.

IBCS will continue to evolve to meet the challenge of new threats as well. Already we are investing in technologies that will incorporate artificial intelligence, machine learning, more automation and designs for a smaller footprint and reduced form factor that will make the capability more mobile and survivable. Nations that already employ IBCS will be able to take advantage of these capabilities as they evolve. In essence, IBCS is a system that can adapt and grow with the future.

2.7 CONCLUSION

At Northrop Grumman, we are driven by a common sense of purpose: our technology helps protect sovereignty and secure freedoms for U.S. citizens and our allies.

IBCS acquisition by the U.S. Army and by U.S. allies enables Coalition Force interoperability and the true network integration required to defeat an ever evolving and complex threat.

IBCS serves as the backbone to our vision of the future as one connected battlespace—one that includes data from the depths of ocean to the vastness of cyber and space.

Through its network-enabled, modular, open and scalable architecture, IBCS gives warfighters capabilities they never had before by fusing sensor data for a single actionable picture of the full battlespace. This ready-now capability gives warfighters more time to make decisions on how best to defeat threats and is a foundational element for enabling joint and coalition, multi-domain operations.

IBCS is a truly transformational C2 capability for IAMD and an enabler for multi-domain integration for the U.S. and its strategic allies. The aviator in me is proud that Northrop Grumman, working alongside the U.S. and our allied partners, is putting the warfighter first in all we do and ensuring they make it home every time.

DATA AVAILABILITY STATEMENT

Data sharing is not applicable to this chapter as no new data were created or analyzed in this study.

FUNDING

This work received no specific grant from any funding agency in the public, commercial, or not-for-profit sectors.

3 Advanced Fluid Dynamics Education and Research in Strategic Technology Domains at the Von Karman Institute for Fluid Dynamics

Peter A. Grognard
Von Karman Institute for Fluid Dynamics,
Sint-Genesius-Rode, Belgium

CONTENTS

3.1 Introduction 19
3.2 Departments at VKI 19
3.3 Educational Programmes at VKI 21
  3.3.1 Research Master in Fluid Dynamics 21
  3.3.2 PhD Programme 22
  3.3.3 Short Training Programme 22
  3.3.4 Master Thesis Programme 22
  3.3.5 Postdoc Programme 22
3.4 Research Expert Groups (REGs) 22
  3.4.1 Rarefied and Plasma Flows 22
  3.4.2 Aerothermochemistry and Heat Transfer 23
  3.4.3 Aerothermodynamics and Heat Transfer 23
  3.4.4 Propellant Management 24
  3.4.5 Aeronautics and Low-Speed Aerodynamics 25
  3.4.6 Aeroacoustics 25
  3.4.7 Shape Optimization 26
  3.4.8 Hypersonic Propulsion 27
  3.4.9 Compressor Aerodynamics 27
  3.4.10 Turbine Aerodynamics 28

DOI: 10.1201/9781003739234-4

3.4.11 Liquid Metals and Industrial Flows 28
3.4.12 Thermal Management and Process Safety 29
3.4.13 Wind Energy and Environmental Flows 29
3.4.14 Machine Learning for Fluid Systems (ML4F) 29
3.5 Research User Groups (RUGs) 30
3.5.1 Instrumentation User Group 30
3.5.2 Computational Fluid Dynamics and Numerical Methods User Group31
3.6 Involvement in AVT and Sci Panels 32

3.1 INTRODUCTION

Founded in 1956 by Professor Theodore von Kármán, the Institute began as the Belgian-American Training Centre for Experimental Aerodynamics (TCEA). It was dedicated to training and research in high-speed aerodynamics, welcoming young engineers and scientists from North Atlantic Treaty Organization (NATO) countries. The goal was to maintain top-tier expertise in aerodynamics within NATO countries. In 1963, the Institute was renamed the von Karman Institute for Fluid Dynamics (VKI).

VKI is financed by 16 NATO countries: Belgium, Bulgaria, Czech Republic, France, Germany, Greece, Hungary, Iceland, Italy, Luxembourg, Norway, Portugal, Romania, Turkey, The Netherlands, and the United States.

The Institute is a world leader in fluid dynamics education and research, specializing in aeronautics, aerospace, turbomachinery, propulsion, and environmental and industrial processes. VKI excels in studying complex dynamic flows through both numerical simulations and extensive experimental research. Success hinges on integrating research into training, linking basic and applied research, and developing programmes aligned with industry needs and international bodies like NATO, the European Union (EU), and European Space Agency (ESA).

The Institute has some of the most important and largest wind tunnel test facilities in the world. VKI operates 48 testing facilities and wind tunnels, which allows VKI to study complex flows ranging from subsonic, over transonic and supersonic to hypersonic flows, with speeds up to Mach 14 as well as plasma facilities such as the well-known "Plasmatron," water tunnel, turbomachinery facilities, vehicle facility, wind gallery, water spray, aeroacoustics and solid propulsion facilities, as well as industrial test rigs.

3.2 DEPARTMENTS AT VKI

In the Aeronautics and Aerospace Department, training and research are conducted for flight vehicles in applications ranging from low speed to the hypersonic regime. Significant experimental and numerical research activity on the aerothermodynamics of space vehicle entry into planetary atmospheres is supported by space and governmental agencies, such as the ESA and the US Air Force Office of Scientific Research (AFOSR), the European Commission (EC) and with a clear link to the vision of the

NATO Science & Technology Organization (NATO STO). Moreover, the activity on cryogenic flows, including hydrogen, is significantly growing with the support of ESA through ESA General Support and Technology Programme (GSTP) and ESA Future Launchers Preparatory Programme (FLPP) projects. Simultaneously, the activities related to subsonic and transonic aerodynamics have been significantly reinforced, through the creation of a dedicated research expert group (REG) "Aeronautics and Low-Speed Aerodynamics (ALSA)," which will gather under its umbrella the research related to Unmanned Air Systems, conventional and disruptive aircraft architectures aimed at improved efficiency. This REG will work in tight collaboration with the REG "Aeroacoustics" for those applications concerned with noise footprint.

In the Environmental and Applied Fluid Dynamics Department, training and research have continued in fields of long-standing interests such as industrial and safety processes, liquid metals in nuclear engineering, the mechanism of turbulence, multiphase and cryogenic flows, aero-acoustics, and wind engineering, the latter related to wind effect on large container in harbor environment, offshore wind energy, and deicing technology. Especially, the activities related to offshore weather prediction and the extreme wind loading and rain erosion of offshore wind turbines and farm wake dynamics are growing. The department is involved as well in international collaborative research on heat exchangers for aeronautics funded by the Belgian Walloon Region, and in the design of Small Modular Reactors (SMRs) together with the Belgian Nuclear Research Centre SCK-CEN.

The Turbomachinery and Propulsion Department has focused its training and research activities in the areas of hypersonic propulsion, aerothermodynamics of high pressure turbines, aerodynamics of boosters, low pressure compressors and turbines, turbochargers, intake performance, design and optimization techniques, and advanced instrumentation technologies for turbomachinery and propulsion components. The European Defence Fund (EDF) programme NEUMANN (Novel Energy and propUlsion systeMs for Air domiNaNce) had the official kick-off meeting in February 2023. VKI contributes with experimental and numerical research targeted to compressor inlet distortion, system optimization, and turbine technologies for next-generation fighter aircrafts. The emerging research field for novel propulsion and energy conversion systems has been advanced in the framework of the Walloon Innovations for Green Skies (WINGS) programme with exploration of approaches towards zero-carbon propulsion technologies. Milestones of phase 1 were successfully achieved, allowing to proceed now with the second and final phase of the programme. With the formerly appointed adjunct professors in Aeroelasticity, further progress in Fluid-Structure interaction in propulsion components was made within the WINGS programme. With the emergent quantum technologies, hybrid quantum computing algorithms are explored for their long-term potential on the simulation and (design) optimization of turbomachinery and propulsion components.

All research engineers, faculty members, and students are part of at least one research expertise group and one user group. The Institute has established a good balance between the necessary academic and research freedom in the spirit of VKI for self-development and exploration and the necessity for accountability on contract research. VKI's comprehensive fluid dynamics expertise is divided into 14 REGs and 2 user groups (RUGs).

The research expertise groups are:

1. Rarefied and Plasma Flows;
2. Aerothermochemistry and Heat Transfer;
3. Aerothermodynamics and Heat Transfer;
4. Propellant Management;
5. ALSA;
6. Aeroacoustics;
7. Shape Optimization;
8. Hypersonic Propulsion;
9. Compressor Aerodynamics and Heat Transfer;
10. Turbine Aerodynamics and Heat Transfer;
11. Liquid Metals and Industrials Flows;
12. Thermal Management and Process Safety;
13. Wind Energy and Environmental Flows;
14. Machine Learning for Fluid Systems.

The two research user groups are:

1. Instrumentation;
2. Computational Fluid Dynamics and Numerical Methods.

These two groups are described further down below.

3.3 EDUCATIONAL PROGRAMMES AT VKI

The Von Karman Institute offers a comprehensive portfolio of educational programmes:

1. Research Master in Fluid Dynamics;
2. PhD Programme;
3. Short Training Programme;
4. Master Thesis Programme;
5. Postdoc Programme.

These programmes are briefly described hereunder.

3.3.1 RESEARCH MASTER IN FLUID DYNAMICS

The main programme offered by the von Karman Institute is the annual Research Master at the level "Master after Master." This programme focusing on teaching research by doing research complements the master programmes of universities with a profound research-oriented specialization. This programme is accredited by the Nederlands-Vlaamse Accreditatie Organisatie (NVAO). The Research Master programme is widely recognized by universities and industries all over Europe and beyond.

3.3.2 PHD PROGRAMME

Candidates typically enter this programme after completing the research master in fluid dynamics. The PhD thesis may be presented at any Belgian university or at a university abroad leading to a degree of doctor in Engineering sciences. The benefits for the PhD candidate are considerable, allowing them to perform research based on the outstanding facilities and unique experience available at the VKI.

3.3.3 SHORT TRAINING PROGRAMME

Qualified undergraduate students from universities or visiting PhD may become acquainted with the work of the fluid dynamicist through participation in a research project at the VKI. The duration of the programme is typically three to six months.

3.3.4 MASTER THESIS PROGRAMME

University students enrolled in an engineering, industrial engineering or physics B.S. or M.S. programme may conduct their final year project at a laboratory of the von Karman Institute.

3.3.5 POSTDOC PROGRAMME

The post-doctoral research programme allows selected top-level researchers to pursue their research after the PhD, taking advantage of the unique research facilities of VKI.

3.4 RESEARCH EXPERT GROUPS (REGS)

3.4.1 RAREFIED AND PLASMA FLOWS

1. *On-Ground Qualification of Intake-Collector System for Airbreathing EP Mission (RAM-EP)*—In the frame of a GSTP programme started in 2021, VKI has developed an optimal design for RAM-EP platform, ensuring an adequate transmission of particles to an electric thruster while minimizing drag. On-ground qualification of an intake-collector engineering model has been completed successfully in relevant Very Low Earth Orbit (VLEO) environments, demonstrating the essential functions of the system.
2. *Magnetohydrodynamic Enhanced Entry System for Space Transportation (MEESST)*—The ionization of the gas surrounding a vehicle flying in hypersonic regime can create a radio frequency blackout and prevent any communication with satellites and ground stations. This is a critical problem for telemetry and system control. In the European framework of H2020 Space calls, the VKI joined forces with other European countries that aim at developing a new technology for blackout mitigation (Fig. 3.1).

Figure 3.1 Integration of the MHD-enhanced probe shielding device in the VKI Plasmatron facility.

3.4.2 AEROTHERMOCHEMISTRY AND HEAT TRANSFER

The research group focus on aerothermochemistry is strongly involved in ESA and EU projects and further develops its expertise for reacting flow and their interaction with thermal protection material. This activity spans from diffusion-reaction process at the surface towards the full demise of the space debris material with all the complex physico-chemical phenomena involved.

1. *Plasma Flow Characterization and Testing Conditions Definition*—Defining accurate testing conditions is a critical aspect of Thermal Protection System (TPS) design, ensuring the reliability of material performance under re-entry scenarios. Last year, significant progress was made in plasma flow characterization, driven by the collaborative efforts of three PhD researchers exploring complementary approaches.

3.4.3 AEROTHERMODYNAMICS AND HEAT TRANSFER

1. *Physics of Nozzle Expansion Flows*—The accurate free-stream flow characterization in all ground facilities is a fundamental requirement for a relevant analysis of wind tunnel data. The flow characterization in rapid flow expansions, as encountered in the nozzle of supersonic and hypersonic wind tunnels, proves to be challenging and "anomalies" have often been blamed onto non-equilibrium phenomena. Intrusive measurements have been im-

plemented in the VKI Longshot gun tunnel, within both contoured and conical nozzle configurations.

2. *Improving Tools for the Design of Planar Supersonic and Hypersonic Nozzles*—The HYpersonic Plasma and Nonequilibrium Shock Environment (HYPNOSE) inviscid nozzle design code has been successfully chained with the DEKAF boundary layer code, enabling the efficient design of axisymmetric supersonic and hypersonic laminar nozzles, and providing an accurate access to local flow quantities, both in the inviscid core flow, and within the boundary layer.
3. *Development of Free-Flight Aerodynamic Measurement Technique*—An advanced measurement technique has been developed for the Longshot wind tunnel, enabling the characterization of the fragmentation process of re-entry bodies upon their re-entry through the atmosphere.

3.4.4 PROPELLANT MANAGEMENT

The projects of this research expertise group are mostly related to (nonreactive) two-phase flows and related heat transfer phenomena in fuel systems for space and terrestrial applications and are described hereunder. The feasibility phase for the VKI Hydrogen Test Facility (VKHyLab) has been successfully completed in accordance with approval by the Belgium Ministerial Council. The VKI research groups are naturally included in the main users of the test facility, to extend their current experimental achievements to the real fuel, at increased test scale. Today, the Engineering Phase is ongoing towards the formalization of collaboration agreements with key users, the concretization of the co-financing, and the launch of the Procurement phases. Construction and operation are planned for the 2026–2027 timeframe.

The Methane Production and Transfer Challenge (MIRAGE) de-risking study (ESA-DERISK framework) was developed in 2021, as the first step towards the extension of the VKI cryogenic research capabilities to liquid Methane, a promising green fuel candidate for space applications. The following phase—MIRAGE PHASE 1—is now progressing to undertake the follow-on steps towards the construction of the MIRAGE Platform.

CRYOSYST project (ESA Technology Development Programme coordinated by Empresarios Agrupados) aims to enhance the capabilities of the ECOSIMPRO library European Space Propulsion System Simulation (ESPSS) toolkit by developing and implementing an interface heat and mass transfer correlation to model evaporation and condensation phenomena in cryogenic fuel tanks during sloshing events. Currently, the ESPSS toolkit does not account for the effects of sloshing on phase-change phenomena, limiting its applicability in such scenarios. The CRYOSYST project addresses this gap by combining dedicated experimental tests with cryogenic fluids with Computational Fluid Dynamics (CFD) simulation. The experimental data provide insights into the underlying phenomena and serve to calibrate the CFD model, which, in turn, offers a comprehensive representation of the thermodynamic variables throughout the domain. The calibrated CFD model will be used to generate a database, which, with the aid of machine learning algorithms, will enable the derivation and calibration of a heat and mass transfer correlation for reduced-order models. This correlation will then be implemented into the ESPSS libraries by Empresarios Agrupados, after which its accuracy will be rigorously tested. Two PhD worked on the project and two conference papers have been submitted to date.

The Hydrogen Aircraft Sloshing Technical Advancements (HASTA) activity (EU HORIZON coordinated by Universidad Politécnica de Madrid (UPM)) has been kicked-off in September 2024 at VKI. The project aims to develop technical guidelines and design criteria for LH2 tanks, adhering to European Union Aviation Safety Agency (EASA) CS-25 regulations. In the framework of the aviation sector's decarbonization, the Hydrogen Pilot STorage for large Ecosystem Replication (HYPSTER) activity (Skywin project coordinated by Safran Aero Booster) focuses on the heat exchanger for the next-generation H2-fueled aircraft. Within HYPSTER, VKI supports the design of the heat exchanger and its experimental testing with substitute fluids.

Be-HyFE aims to build a Belgian hydrogen homebase of academic knowledge and industrial expertise, funded by the federal Energy Transition Fund. VKI hosts one of the 16 network PhDs, in collaboration with Ghent University.

3.4.5 AERONAUTICS AND LOW-SPEED AERODYNAMICS

The ALSA REG has been created in 2024 in order to better address emerging challenges in the aeronautical industry, as well as in other sectors sharing similar issues in terms of performance, control, or resilience to adverse atmospheric conditions, e.g. wind turbine blade flows. On key driver in the aeronautical sector is the advent of novel aircraft architectures and propulsion concepts such as Boundary Layer Ingestion or Distributed Electric Propulsion. Supersonic aircraft propulsion has been researched in the framework of the H2020 project MORE&LESS (MDO and Regulations for Low-boom and Environmentally Sustainable Supersonic aviation), with a focus on the shock-cell pattern developing in the jet plume of a hybrid turbojet/RAMJET engine.

In the transonic and low-supersonic ($Ma < 2$) regimes, the REG is also helping the space launcher sector. This activity started with the ESA Launcher Payload Fairing Transparency Preliminary Assessment (FAIRTRANS) project, is now followed by the ESA MATTHIAS project, aimed at developing lighter and more performant thermal and acoustic insulating materials.

Research on ice accretion on aircraft and wind turbine leading edges continues at VKI since the modification of the Cold Wind Tunnel CWT-1 to an icing wind tunnel.

3.4.6 AEROACOUSTICS

The Aeroacoustics REG has pursued its acoustic multiport research activities for the noise and vibration emitted by ducted singularities, through research contracts with industrial partners. The application is the detection and quantification of leakages in pipeline systems found in gas distribution networks, nuclear plants, etc., for example caused by faulty valves.

Research on noise mitigation strategies is pursued in the framework of the H2020 RIA Project ENODISE ("ENabling Optimized DISruptivE airframe-propulsion integration concepts"), coordinated by Prof. Christophe Schram.

VKI has also expanded the initial scope of its activities in the ENODISE project by initiating a research programme on multicopter drones. Experimental campaigns have been carried out in the Jet Aeroacoustic Facility for Aeronautical and Aerospace Research (JAFAAR) anechoic laboratory, in free field and in the premises of the ID2MOVE drone-port. Also on jet noise, the H2020 RIA project MORE&LESS considers the noise emissions by the propulsion system of future supersonic aircraft. The aeroacoustics group will carry out experimental investigations in JAFAR, to be cross-validated with the numerical simulations planned by the MORE&LESS partners and propose low-order models.

Also in the high-speed flow range, the group has been extending the research that was initiated in the ESA GSTP project FAIRTRANS, with a new ESA GSTP project MATTHIAS kicked off in 2023, for the detailed experimental investigation of aero-vibro-acoustic coupling mechanisms relevant to payload fairing design, with a focus on the acoustic performance of thermal shielding samples.

Last but not least, VKI has obtained the organization of the 32nd American Institute of Aeronautics and Astronautics (AIAA)/Council of European Aerospace Societies (CEAS) Aeroacoustics Conference, which will take place in Brussels in 2026. This will be the first edition of this prestigious conference—the main international event for the aeroacoustics community—to be organized in Belgium.

3.4.7 SHAPE OPTIMIZATION

The shape optimization REG focuses primarily on the design optimization of turbomachinery components and has developed software that enables the design optimization of various components in a multidisciplinary environment. Since several years, the group has extended the optimization tool with gradient-based methods, which allow for a faster convergence to the optimal solution within a richer design space at a lower computational effort.

The software developed by VKI has been extensively used in the power2olefines project, where non-conventional turbomachinery is used to provide the necessary heat for so-called cracking, a chemical reaction to convert carbohydrates into light olefines. A pilot prototype design has been developed at small scale to perform reactive experiments and validate the concept. Due to the small scale, a drive turbine had to be designed as substitute for the electric engine that drives the reactor in real scale. The drive turbine has been optimized to reduce the operation cost of the pilot reactor. Currently, the reactor is being manufactured, and the tests are expected to start early next year.

The optimization software has been further applied in industrial projects for the design of radial turbines in cryogenic conditions, as well as the application of hub endwall contouring for axial turbines. A continuous effort is made to apply the tool to industrial problems.

3.4.8 HYPERSONIC PROPULSION

1. *Hypersonic & Supersonic Propulsion Research*—In the framework of multiple collaborative projects, the Hypersonic Propulsion and Combustion Research Expertise Group conducts cutting edge numerical and experimental research as well as system level modelling of hypersonic and supersonic propulsion systems. The EU-Horizon 2020 sponsored MORE&LESS project has been under execution since January 2021 and will continue until April 2025. The project aims at investigating various architectures of sustainable supersonic civilian aircrafts to operate at various flight Mach numbers, between 2 and 5, by considering their environmental footprint in terms of pollutant emissions and noise as well as long-term climate impact and consequently reducing the environmental impact of high-speed aviation through multi-disciplinary optimization of aircraft concepts.
2. *Emerging Combustion Technologies Research*—The project Additively Printed Engine (APE) was awarded funding by the NATO Science for Peace and Security office (NATO-SPS) in October 2021 to investigate a new class of small-size turbine engines printed through a single shot additive manufactured process to minimize the unit cost. The NATO-SPS-APE project not only is the first SPS project that VKI was granted but also will advance the porous media combustion modelling research at the Institute.
3. *EDF project NEUMANN*—coordinated by GE Avio aims at development of a novel propulsion system and its critical components to power next-generation European fighter aircraft concept. The project involves 38 partners including leaders of aerospace industry and academia from 14 European nations.

3.4.9 COMPRESSOR AERODYNAMICS

The aerodynamics of compressors remains an important research activity at the VKI. These investigations address roughness, transition, separation, and boundary layer control.

High-speed/high-turning low-pressure compressor stator and rotor profiles have been designed within the WINGS project funded by the Walloon Region. The validation campaign has been finalized in the fourth quarter of 2023 for stators while the high-speed rotors campaign is ongoing. The experimental activity has been supported by the novel design of two tests sections.

The compressor research group relies on the collaboration with several international institutions such as Ecole Centrale de Lyon, Imperial College, Cambridge

Figure 3.2 The research high-speed low-pressure turbine stage SPLEEN successfully tested in the VKI CT-3 turbine facility in 2023.

University—Whittle Laboratory, Politecnico di Milano, University of Liege and the Turbomachinery Laboratory of the University of Notre Dame.

3.4.10 TURBINE AERODYNAMICS

As for compressors, the aerodynamics of turbines remains an important research activity, too. Similar as for compressors, these investigations address roughness, transition, separation, and boundary layer control.

Within the Horizon Europe HE-ART project, VKI oversees generating a highly detailed experimental dataset on the effect of high freestream turbulence on turbomachinery blades. The Cleansky2 project SPLEEN deals with the investigation of secondary and leakage flows in low-pressure turbines (Fig. 3.2).

The EDF-funded project NEUMANN kicked-off in February 2023. VKI is responsible for the experimental characterization of novel technologies and designs for the transonic turbines that will power the sixth-generation fighter aircraft.

3.4.11 LIQUID METALS AND INDUSTRIAL FLOWS

In the frame of VKI's long-term collaboration with the Belgian Nuclear Research Centre (SCK-CEN), a new five-year Memorandum of Understanding was signed between the two institutions to support the design of the lead-cooled SMR lead-cooled Small Modular Reactor (LEANDREA) (2024–2029). The thermal hydraulic research focuses on design evaluations, operating conditions and design safety and assessments, including passive cooling devices.

The VKI is involved in safety analysis support by providing models addressing possible safety-related scenarios like for example ex-vessel and in-vessel decay-heat removal studies, conceptual assessment of passive cooling by water injection, diaphragm break, or fuel debris accumulation.

3.4.12 THERMAL MANAGEMENT AND PROCESS SAFETY

This new REG was created to merge the growing activities in heat exchangers in aeronautics, with the existing research related to process safety.

Aerothermal measurements of compact heat exchanger by additive manufacturing is an activity that encompasses two main projects: HYPSTER, and PIT HEX, part of the WINGS.

The HYPSTER project, shared with the propellant management REG, oversees helping in the design and of the experimental characterization of heat exchanger prototypes that will be part of future hydrogen airplanes.

The second project related to heat exchangers in additive manufacturing is the PIT WINGS project, investigating air/oil aircraft engine heat exchangers extracting thermal power from oil to air.

3.4.13 WIND ENERGY AND ENVIRONMENTAL FLOWS

CFD of atmospheric flows is of great interest for wind engineering and wind energy applications, including wind loading, pedestrian wind comfort, offshore wind energy, and pollutant dispersion.

VKI is part of the ETF project PhairywinD project to support the current and future Belgian offshore wind energy sector. In offshore wind energy, VKI is also leading the SeaFD project related to realistic CFD wind load computations for offshore wind turbines to find strategies to mitigate the effect of extreme weather conditions.

The Horizon Europe project DTWO marks the first VKI project in offshore wind energy at the European level. The project has a multi-disciplinary scope, in which a federated digital twin platform for offshore wind farms will be built.

The platform will account for aerodynamic interactions between offshore wind turbines, both through mutual wake interactions and with the global weather system through the EU Destination Earth initiative.

3.4.14 MACHINE LEARNING FOR FLUID SYSTEMS (ML4F)

The integration of machine learning methods in fluid mechanics continues to grow and open new frontiers. Combining first principles and machine learning offers new avenues to improve predictions and control design, identify patterns, reduce computational costs, enhance measurement techniques, and more generally, get deeper insights into complex fluid phenomena.

The VKI is at the forefront of both research and education on machine learning for fluid dynamics. The first lecture series on the topic, organized in 2020 and attracting more than 200 international participants, has been published in an edited book by

Cambridge University Press. The second edition, in 2024, reached 240 participants. In parallel, a new course titled "Hands on Machine Learning" is now approaching its fifth edition, attracting more than 100 participants every year. Together with the VKI Research Master students, these courses have been followed by more than 1000 participants.

Currently, new courses and a new curriculum in the Research Master on "Machine Learning for Fluid Dynamics" have been introduced. Given the prominence of machine learning and digital twinning technologies, the ML4F group has developed a new set of courses that will constitute a new specialization on Machine Learning for Fluid Dynamics within VKI's Research Master programme.

Starting from this year, in addition to the classic "experimental fluid mechanics" and "numerical fluid mechanics" curriculum, the Research Master's programme will offer a new specialization in "Machine Learning for Fluid Mechanics," featuring four dedicated courses and laboratory activities focused on data assimilation, digital twinning, and adaptive control of fluid systems. With the ever-growing need for researchers trained in both fluid mechanics and machine learning algorithms, the new curriculum offers a unique opportunity to gain experience and conduct research at the forefront of machine for fluid dynamics!

A major European Research Council (ERC) Project has been granted to the ML4F group. Dr. M. A. Mendez, head of the ML4F group, has been awarded the prestigious ERC Starting Grant for his project "Re-Twist: Reinforcement Learning Systems." This grant, part of the EU's Horizon Europe programme, supports early-career researchers in establishing independent teams and pursuing innovative research. The Re-Twist project aims to bring together several areas of research on which we have been active in the last three years at the von Karman Institute. These are:

1. real-time data assimilation and modelling of engineering systems;
2. machine learning–based surrogate modelling;
3. machine learning–based control methods.

These fields are "twisted" under one single goal: developing algorithms for real time.

3.5 RESEARCH USER GROUPS (RUGs)

The two research user groups are:

1. Instrumentation (Research User Group);
2. Computational Fluid Dynamics and Numerical Methods (Research User Group).

These groups are described here below.

3.5.1 INSTRUMENTATION USER GROUP

Driven by the continuous development of key technologies such as imaging sensors and laser illumination sources, measurement chains are continuously increasing in size, cost, and complexity. This is particularly true for velocimetry techniques, with modern tomographic Particle Image Velocimetry (PIV) systems capable of providing 3D velocity measurements at a frequency of several kilohertz on volumes as large as a cubic meter. The terabytes of velocity data provided by each measurement brings challenges and opportunities both at the processing and at the post-processing phases. Instrumentation development and application in turbomachinery has always been of major interest to VKI. VKI continues to design, build, and operate total pressure probes that can withstand the harsh gas turbine (including the combustion chamber) and aero-engine or industrial gas turbine environments. Developments with some important industrial partners are still going on. In parallel, several initiatives have been ongoing to apply optical techniques to measure the velocity field in turbomachinery types of flow using endoscopic probes to introduce the laser sheet.

Significant efforts have been made in recent months to provide top-quality calibration certification for pressure/temperature rakes used in aero-engines. The close relations built over the past decades with several major industrial partners continue to be further expanded.

The numerical prediction of the complex flow field in a pressure/temperature measurement head has been continued with the aim of optimizing these probes.

Substantial efforts have also been addressed to the measurement of turbulence in harsh flows (compressible, low density, and high temperature) by means of hot wires and fast response pressure probes.

The development of experimental techniques to measure the turbulent heat flux in liquid metals has started in collaboration with Belgian's nuclear research centre SCK-CEN.

3.5.2 COMPUTATIONAL FLUID DYNAMICS AND NUMERICAL METHODS USER GROUP

The Numerical Methods User Group bundles activities across VKI's departments and REGs. Its main purpose is to provide a forum for activities related to computational methods, especially in the realm of physics-based simulation (CFD, Direct Simulation Monte Carlo (DSMC), etc.). Focus areas of the user group include code development and management, formulating CFD strategy, and coordinating research in numerical methods.

In addition to coordinating active research, the user group functions as a forum for discussion of novel research and such technology, which within application areas traditionally considered by VKI is not considered standard.

Research highlights of the Numerical Methods User Group include:

1. high-order discontinuous Galerkin method for inductively coupled plasma;
2. high-order simulations of noise-generation in hypersonic boundary layers;
3. anisotropic metric-based adaptation for time-dependent high-order methods.

3.6 INVOLVEMENT IN AVT AND SCI PANELS

Over the years, the VKI has established a strong relationship with the Applied Vehicle Technology (AVT) and Science (SCI) panels of NATO's Science and Technology Organization (STO), and continues its support in activities to further advanced propulsion systems:

1. *NATO Science and Peace and Security (SPS)*—Dr. Peter Simkens, Business Development Manager at VKI, is appointed as official expert to the SPS Programme's Independent Scientific Evaluation Group (ISEG) for the period 2024–2027.
2. *Collaboration Support Office (CSO)*—VKI and CSO met twice in June and September 2024 to discuss the interactions of the Collaboration Support Office with the Lecture Series team, fostering better engagement. Additionally, it is also a key opportunity to review the procedures established for the conduct of co-sponsored Lecture Series (LS).
3. *First STO/VKI Lecture Series on Advanced CFD Methods for Hypersonic Flows*—A first lecture Series AVT-358 on "Advanced computational fluid dynamics methods for hypersonic flows" was held on site in March 2024 and gathered 64 participants from 15 NATO countries. A second lecture series will be organized in July 2025 at Stanford University.
4. *Second STO/VKI Lecture Series on Quantum Computing*—Prof. Frank Eulitz and Dr. Wim Munters have developed a new Research Lecture Series "Introduction to Quantum Computing in Fluid Dynamics" with STO-AVT-377 which was successfully held in a hybrid mode in May 2022 with a total of 134 registrants and with further extension in July 2024 with a total of 260 registrations from 20 NATO countries. Reflecting on the emergent character of Quantum Computing research and development both the format and the content were further advanced for the second LS. Aside of further applications and presentations from companies and academic institutions, the lecture series included industry impulse sessions on four lecturing days shedding light on the advancements in noise detection, correction, and mitigation and engineering application.
5. *SPS Projects*—The project APE was awarded funding by the NATO Science for Peace and Security office (NATO SPS) in October 2021 to investigate a novel class of turbine engines printed through a single shot additive manufactured process for three years. VKI is the coordinator of the consortium composed of Technion (Israel) and Izmir Katip Celebi University (Turkey). The project ended successfully in September 2024.

DATA AVAILABILITY STATEMENT

Data sharing is not applicable to this chapter as no new data were created or analysed in this study.

FUNDING

This work received no specific grant from any funding agency in the public, commercial, or not-for-profit sectors.

4 The Issues of Modeling and Simulation of IAMD

Josef Časar
Department of Air Defence, Faculty of Military Technology,
University of Defence,
Brno, Czech Republic

CONTENTS

4.1 Introduction 34
4.2 Types of Simulators 35
4.2.1 Live Simulation 35
4.2.2 Virtual Simulation 36
4.2.3 Constructive Simulation 36
4.2.4 Tactical Simulators 37
4.3 Possibilities of Using Simulators 37
4.3.1 Level-Specific Simulations 38
4.3.2 Operation-Specific Simulations 38
4.3.3 Capability-Specific Simulations 39
4.3.4 Support of Cyclic and Repetitive Processes 39
4.4 Performance Requirements 39
4.5 Examples of Simulators 40
4.5.1 Extended Air Defense Simulation 40
4.5.2 Interactive Simulation Package 41
4.5.3 Virtual Battlespace (VBS) 41
4.5.4 VR-Forces 42
4.6 Potential Applications of Contemporary Technologies 42
4.7 Conclusion 42

4.1 INTRODUCTION

Modeling and simulation (M&S) in the military environment have experienced a significant boom in the last decade. Thanks to the development of information technology and increasing computing power, it is increasingly possible to provide more comprehensive and realistic models not only of military equipment but also of the processes and tactical principles of their use. Increasingly sophisticated models are not only visually superior but they are also more realistic in terms of physical and

DOI: 10.1201/9781003739234-5

tactical behavior. Modern simulators are then able to place these models in a virtual environment, providing users mainly with two basic capabilities. The first is the ability to test and validate hypotheses about the use of forces and assets in a contemporary theater of war. The second is undoubtedly the ability to teach and train personnel in the use and control of equipment and in tactical skills. The field of ground-based air defense (GBAD) is no exception in this respect. Unlike other types of armies, however, GBAD is in a way handicapped by manufacturers and developers because it involves a highly specialized field of application of anti-aircraft missile systems, which are, after all, less numerous in their use in the armies of the world than, for example, those of the air force or conventional ground forces. For this reason, it is not profitable for large development companies to offer already developed environments capable of authentic simulation of GBAD assets, as each nation, including members of the North Atlantic Treaty Organization (NATO), has different types of anti-aircraft missile systems in its arsenal. At the same time, it is necessary to take into account the fact that the capabilities of anti-aircraft missile systems are in the vast majority of cases subject to some level of classification, making the identification of the exact characteristics of these systems very difficult and more or less impossible in a commercial environment. Many users therefore address M&S of GBAD forces and assets independently within their classified defense research projects, making the development of this capability on a global and commercial scale difficult.

This work addresses the issues and aspects of M&S of GBAD assets and forces represented by M&S capabilities including capability modeling and then examples and experiences with specific simulators. The analyses presented merge most of the current requirements in GBAD M&S. The work was preceded by many years of research by the author, who summarizes here all the findings and results of research on the subject in the period 2009–2025.

4.2 TYPES OF SIMULATORS

Currently, three fundamental categories of simulators, in conjunction with their combinations, predominate in the realm of military professional training. From a technical point of view, these systems are classified as live simulation systems, virtual simulators and constructive simulators. The distinguishing characteristics of these simulators primarily lie in the quantity of virtual elements employed and the array of technical and technological systems utilized for their implementation. In terms of utilization, setting aside the technical intricacies, the discussion encompasses the incorporation of tactical simulators.

4.2.1 LIVE SIMULATION

The live simulation represents an exercise in which units on the field utilize real equipment and weaponry with training ammunition. The weapons are equipped with a system simulating the firing of live ammunition. This type of simulation is used to train and simulate combat conditions and possible deployment of units in combat operations.

4.2.2 VIRTUAL SIMULATION

In the virtual simulation, in addition to the effects of ammunition, all technical equipment, weapons, machinery and the environment in which the trainee moves are also modeled and simulated. Orientation and movement in the artificial environment are mediated by three-dimensional displays, and reaction to the situation is enabled by the simulator controls. Depending on the need and purpose of the whole simulation, the controls are chosen to be very faithful to reality or can be simplified in some way.

Depending on the fidelity of the design and layout of the individual parts of the virtual simulator, which depend primarily on the intended use, these simulators are further divided into two types:

1. *First Type Virtual Simulators*—when training on a first type virtual simulator, the emphasis is on the fidelity of the vehicle type. The internal layout of the combat vehicle simulators is identical in detail to the actual vehicles used as far as the controls are concerned. This makes them suitable for training not only in tactics, command and control but also for practicing the control of the elements of a given combat vehicle during shooting and driving training. This allows for the training of drivers, gunners, commanders and swarms in a virtual reality environment. Soldiers will experience controlling vehicle features, monitoring engine specifications and troubleshooting technical issues. They can practice overcoming terrain obstacles, communication within the crew, as well as practicing shooting and tactical situations.
2. *Second Type Virtual Simulators*—the second type of virtual simulator is a simplified version of the real resources. Common computer components are used as control and communication elements. The monitor displays the most necessary devices and a view of the virtual terrain, and the joystick enables control of the basic elements of the technique. In a very short time, one of the established configurations of each type of vehicle can be configured for each exercise according to the requirements of the trainees. Virtual simulator programs are also available to enable dismounted infantry exercises. These simulators are intended for training in tactical training and command and control mainly. Their strength lies in the ability to interface not only with the first type of virtual simulators but also with the constructive simulation system, and it is possible to configure the required number of entities (ground vehicles, air vehicles, etc.) so that the composition and numbers of vehicle for example in convoys match the requirements of the trainers.

4.2.3 CONSTRUCTIVE SIMULATION

The concept of constructive simulation represents technologies and processes in which vehicle crews and individuals are virtualized in addition to the above elements. Constructive simulation is a networked computer strategy game that takes place in real time and can involve a large number of players. However, these simulators are much more sophisticated and also include terrain models with some accuracy, realistic parameters of equipment, weapons and ammunition based on their actual characteristics. The individual details of the combat exercise scenario are prepared according to the given control and combat documentation, thus preparing the exercise participants directly for a specific combat issue. The course itself can be further analyzed by assessing the weaknesses and strengths of individuals or the whole units during and after the simulation itself.

4.2.4 TACTICAL SIMULATORS

The main purpose of the tactical simulator is to enable collective training of specific forces (ground troops, pilots or commanders) in addition to individual training. This type of simulator contains the necessary tools for effective planning and execution of the entire training operation and all its tasks. It is primarily aimed at assessing and improving collective tactical knowledge and skills. To improve the training effect, it should also include evaluation tools for assessment during operations and for final evaluation after operations.

Exercises on tactical simulators are mainly focused on improving the following skills and abilities:

1. information gathering, understanding and general awareness of the tactical situation of own and enemy troops;
2. analysis and quantification of the situation;
3. decision-making process;
4. commanding units, issuing commands to subordinates, and conveying commands from superiors;
5. appropriate use and handling of assigned assets and supplies;
6. reporting the situation to superiors.

4.3 POSSIBILITIES OF USING SIMULATORS

It is almost not possible to involve all the parameters and influences from real world and usually it is even not desirable. A very important parameter and basis for M&S is the environment model. The environment includes everything that is required by the simulated entities. In the case of terrestrial entities, the focus is usually more on terrain modeling, surface type, relief, etc. For airborne forces, clouds, wind or different temperatures at different altitudes are important. In the case of maritime entities, it may be sea currents, salinity or also the state of the surface with respect to wave height. The ocean as such has its own climate, which is as important for submarines as the climate of the atmosphere is for aircraft. In the case of modeling space vehicles, it is again necessary to take into account the so-called space weather. Thus, whatever is perceived as important for the defense domain should be included, not simplified or abstracted. If a particular environmental attribute is required, it must be captured, modeled and embedded with the required accuracy in some part of the simulation.

Along with the environment, other parameters and required properties of the simulators must be considered according to different purposes of specific simulations. Furthermore, we can define the division of simulators according to level of military operation, type of operation, required capability or the purpose of use.

4.3.1 LEVEL-SPECIFIC SIMULATIONS

According to the AJP 3.0—Joint Operations, there are no exact boundaries between the levels of warfare. But as well as they help the commander to design and synchronize the operations, allocate resources and assign tasks to appropriate level of command, and their clarification is also essential when preparing a simulation or designing a simulator. When simulating specific military operation level, it is necessary to think about the appropriate strategic, operational or tactical nature of objective, mission or task which could relate to following issues [1]:

1. echelon of command and control (including possible overlapping);
2. areas of responsibility;
3. size of units;
4. types of equipment.

4.3.2 OPERATION-SPECIFIC SIMULATIONS

According to the type of operation, the simulations and the range of simulated units and elements may vary by an order of magnitude—from one purpose simulations of specific High-Value Asset (HVA) defense through simulation of a specific kinds and parts of the forces to complex scenarios limited only by the type of area. Another operation-specific simulations can include:

1. HVA defense;
2. maneuvering forces defense;
3. defense in urban areas;
4. defense in suburban areas;
5. defense in rural areas.

Finally, there may be combinations of different types of operations with the main type of force that may be negatively affected by the absence or limited functionality of GBAD being fighter pilots, who need to train against potential enemy GBAD to gain the necessary habits and understanding of the physical capabilities and capabilities of GBAD itself. Without simulators with realistic GBAD models, only limited training can be conducted, consisting mostly in tactics to siege enemy GBAD forces. However, it is not possible to gain knowledge of realistic capabilities and thus imminent danger.

4.3.3 CAPABILITY-SPECIFIC SIMULATIONS

In the context of acquisition processes as well as when defining, achieving and declaring achieved capabilities within an alliance, the functionality to simulate a specific capability (Counter-Unmanned Aerial Systems (C-UAS), Counter-Rocket, Artillery and Mortar (C-RAM), etc.) or a combination (layered AD) thereof seems to be very useful.

4.3.4 SUPPORT OF CYCLIC AND REPETITIVE PROCESSES

The simulations can be used as a support tool for different military processes where they can represent significant reduction of financial, time and personnel requirements. Moreover, the importance is even greater in case of cyclic or repetitive processes:

1. C2 process support;
2. planning process support;
3. capability development process support;
4. acquisition process support.

4.4 PERFORMANCE REQUIREMENTS

Due to the speed of objects in airspace, fast information processing is very important in integrated air and missile defense (IAMD). However, due to the necessary complexity of the simulations, it can often be a problem to process large amounts of data in this short time. If we want to cover the whole range of IAMD from the surface (UAS) to space (in the case of ballistic missiles), we are interested in a large part of the environment, and with a lack of computational power or transfer capacity, it is necessary to establish possible simplifications (create a model and neglect less important parameters).

M&S of GBAD is a very complex process. Although there are many products that are capable of simulating GBAD activities to a limited extent, their characteristics and features are very limited, without deeper implementation of tactical capabilities and principles. This fact reduces the applicability of these products and does not allow the user to study, train or evaluate GBAD procedures and capabilities as a whole. Furthermore, it is currently also not possible to simulate the functionality tactical command and control centers of Air Force (AF) that would allow the automatic control of the operation of modeled combat entities to ensure compliance with tactical rules of engagement and coordination of the combat effort of forces and assets. This absence is overwhelmingly replaced by the presence of "live" operators or instructors who interact with the simulation or controlled entities and thus substitute for the activities of the more complex model of the operational-tactical element (center). However, in terms of the demands placed on the scope and purpose of simulation technologies, this trend is unsustainable due to increasing demands on simulation capabilities. Increasingly, it is necessary to simulate not only basic scenarios, such as the simulation of a single air battle or a time-limited air operation, but also more complex operational-tactical scenarios, such as the ability to counter different types of air attack with different variants of available forces and assets, with different command hierarchy settings and under different geopolitical conditions (peace, crisis, and war). Thus, it is currently almost impossible to know how a particular system will behave because models of GBAD assets and their capabilities are not available.

When developing or buying a simulator it is necessary to think about above-mentioned and required complexity. It is a well-established fact that the complexity of a simulator is directly proportional to its level of control complexity. Moreover, the higher control complexity places higher demands on the operators and their skills which could lead to the need for an IT specialist instead of just an IAMD Subject-Matter Expert (SME).

On the one hand the complexity can be reduced using single-purpose (single-domain, single-capability, etc.) simulator, but on the other hand a multi-domain and multi-capability simulator can give us better analytical and other results.

Simulators and simulations can be approached in a manner consistent with the description and acquisition of real capabilities. They can be summarized in the following groups:

1. C-UAS;
2. C-RAM;
3. Short-Range Air Defense (SHORAD);
4. Medium-Range Air Defense (MRAD);
5. Long-Range Air Defense (LRAD);
6. Army Organic Air Defense.

However, these core capabilities must always be complemented by supporting capabilities from which the communication plays an essential role. Modern simulation tools allow us to deal with the full range of communication modes, options and parameters in a similar way to real deployments. Examples include data vs voice and classified vs unclassified.

In addition to the aforementioned capabilities, there are others that can be included as needed for a specific simulation. These include modeling, simulation and analysis capabilities for logistics support, including supply and movement.

4.5 EXAMPLES OF SIMULATORS

In this section, the current state-of-the-art analysis briefly introduces the products that enable basic and advanced training of air defense personnel.

4.5.1 EXTENDED AIR DEFENSE SIMULATION

In the area of GBAD, there are also products that are more narrowly specialized in air warfare. Representatives of this category include, for example, Teledyne Brown Engineering, which deals with air and missile defense. Its major products include Extended Air Defense Simulation (EADSIM), a software that enables simulation of combat between airborne and GBAD entities, enabling subsequent analysis of operations and testing of operational planning. However, the product is primarily focused on missile defense and defense using long-range systems, which could be difficult to use with the potential necessity of simulating short or even very short ranges. Despite that a use could be found for analytical purposes or for capability-specific simulations [2].

4.5.2 INTERACTIVE SIMULATION PACKAGE

This simulation tool is developed by the NATO Programming Center (NPC) in Glons, Belgium, and its main purpose is to provide the alliance's air forces with a tool for creating simulated air situations. It is a kind of air target generator in which it is possible to simulate any air situation scenario ranging from conventional air traffic to scenarios of air attack by enemy air forces. However, the output is only a situational air picture in the form of an input to the screens of operations center personnel. The program cannot simulate GBAD assets. Thus, for use by the GBAD, this product can only be used as a support tool for simulated airborne situations.

4.5.3 VIRTUAL BATTLESPACE (VBS)

The simulation environment of Bohemia Interactive company, originally developed by Czech programmers, is probably one of the best available complex tools on the military simulation technology market at present, alongside the product of MAK Technologies. The products are divided into several applications that cater for different military simulation requirements. The main product is the Virtual Battlespace (VBS), currently in version 4, which is a virtual training environment with databases and libraries of more than 14,000 models of equipment, weapons and people.

Other interesting products of this company include the VBS Control Behavior Pack 1 software for simulating the behavior of people (individuals, crowds, etc.), which can be effectively used especially in tactical training of ground units in urbanized environments. The creation of scenarios and the insertion of own modeled entities (e.g., also GBAD assets) into the simulation can be performed in the VBS Simulation SDK development environment. An interesting product increasing the training possibilities is the application for simulation and training of radio communication called VBS Radio Pro [3].

As with other mentioned products, it is not possible to model and simulate more sophisticatedly the capabilities of GBAD assets in training. To a limited extent, it is possible to train tactical procedures that are common to all military units without distinction. However, it is not possible to create scenarios focusing on the tactical activities of specific GBAD forces. The RBS-70 model, as a part of the simulator, can be considered among the most successful achievements of this company in terms of support for simulation of GBAD.

4.5.4 VR-FORCES

Another well-known company that creates robust simulation environment is MAK Technologies, Inc. They are one of the companies that have a very sophisticated simulation environment capable of many functionalities that are being adopted by some of the other big players in the simulation field. The environment emphasizes the ability to link simulations across a wider set of users. The main product is the VR-Forces tool, which enables the integration of workplaces into a wider range of simulations within the High Level Architecture (HLA) Distributed Interactive Simulation (DIS) environment. In simulations it is possible to create complex combat scenarios with a wide set of generated entities (environment, people, technology, etc.) [4]. Other products increase the realism of the simulations by, for example, implementing night vision or infrared visualization, modeling radars and sensors, or modeling the behavior of people (DI-GUI). However, similar to other products, the M&S of GBAD assets and the physical processes of warfare in the form of the process of guiding missiles to targets and the tactical principles of use are not satisfactorily addressed.

Other examples of military/GBAD simulators could include SWORD from MASA Group or Joint Theater Level Simulation - Global Operations (JTLS-GO) from Rolands and Associates under Valkyrie Enterprises, Inc.

4.6 POTENTIAL APPLICATIONS OF CONTEMPORARY TECHNOLOGIES

With the development of modern technologies, whole systems or parts of systems are constantly being replaced by systems with elements of artificial intelligence and other elements from various disciplines. All this leads to a gradual increase in the overall effectiveness of combat operations.

Not only the commander's awareness but also the regular evaluation of information on the course and results of combat activity contributes significantly to the increase of combat effectiveness in the case of planning and conducting combat. Several tools can be used to support the planning process. Almost any of the constructive simulators mentioned, including analytical ones, can be used for this purpose.

4.7 CONCLUSION

The possibility of using a comprehensive simulation environment will be indispensable for the personnel of GBAD forces in the future. Currently, every new anti-aircraft missile system has a simulator or trainer produced by the manufacturer. At best, a fire control simulator. However, we cannot speak of a comprehensive simulation environment in which to analyze and rehearse the deployment of firing systems to study deployment options. What is meant here is primarily the possibility to simulate the deployment of larger or smaller task groups of the AF in the defense of assigned objects or areas. The GBAD has an additional training requirement that current simulation technologies do not usually provide, which is the ability to simulate combat operations in parallel with their own air forces. In order to safely interact in the contemporary theater of war (which is characterized by significant dynamics and high information flow), pilots and airborne personnel need to practice interacting in a common area of responsibility.

Wargaming in conjunction with M&S represents an integral part of any modern military operation planning or analysis process, regardless of forces kind. There are a number of tools used in the security and military domains, including tactical constructive simulators or constructive wargaming platforms, some of which have been presented in this chapter. These tools are often very robust and complex and can be used for a variety of applications involving wargaming or Course of Action (COA) planning. However, their robustness and complexity are also associated with higher user, time and technical implementation requirements and they tend to be more complex to operate. These are the main disadvantages of the aforementioned commercially available solutions. Another disadvantage may be the higher price or the lack of knowledge of the internal structure of these closed systems.

DATA AVAILABILITY STATEMENT

Data sharing is not applicable to this chapter as no new data were created or analyzed in this study.

FUNDING

This work was supported by institutional support from the Ministry of Defense of Czech Republic.

REFERENCES

1. NATO Standard AJP-3(C). *Allied Joint Doctrine for the Conduct of Operations*, February 2019.
2. Extended Air Defense Simulation. 2019. Teledyne Brown Engineering, Huntsville (AL), USA. https://www.tbe.com/missionsystems/eadsim (accessed April 20, 2025).
3. Advanced Air Defense Simulator. 2019. Bohemia Interactive Simulations, Williamtown, Australia. https://bisimulations.com/ (accessed April 20, 2025).
4. VR-Forces. 2019. MAK Technologies, Cambridge (MA), USA. https://www.mak.com/ (accessed April 20, 2025).

5 Robust Resource Allocation for C-UAS Defense
A Reinforcement Learning Approach

Piotr Turek
Faculty of Mechatronics, Armament and Aerospace,
Military University of Technology, Warsaw,
Poland

Mateusz Kaszyński
Faculty of Mechatronics, Armament and Aerospace,
Military University of Technology, Warsaw,
Poland

Jakub Majchrowicz
Faculty of Mechatronics, Armament and Aerospace,
Military University of Technology, Warsaw,
Poland

CONTENTS

5.1 Introduction 45
5.2 The WTA Problem 45
5.3 Simulated Environment 48
5.4 Problem Formulation and Solution Approaches 53
  5.4.1 Problem Formulation 53
  5.4.2 Solution Approaches 54
    5.4.2.1 Integer Linear Programming (ILP) 54
    5.4.2.2 Q-Learning (QL) 55
    5.4.2.3 Proximal Policy Optimization (PPO) 57
5.5 Methods and evaluation 60
5.6 Results 62
  5.6.1 Computation time 62
  5.6.2 Solution quality 63
  5.6.3 Selector-based evaluation 63
  5.6.4 Tactical scenario verification 66

DOI: 10.1201/9781003739234-6

5.7 Conclusions 67

5.1 INTRODUCTION

Observations of the modern battlefield and various armed conflicts around the world indicate that the strikes of the adversary are performed with growing frequency as saturation attacks. In the context of air defense, a saturation attack is a tactic that employs a very large number of aerial threats of different types, such as ballistic missiles, cruise missiles (CM), or unmanned aerial systems (UAS). The main purpose of such an attack is to force the defender to use their limited and valuable defensive resources to the maximum extent and to occupy as many fire control channels as possible. This creates opportunities for other threats to penetrate the defended area and execute successful strikes, which cannot be intercepted due to the saturation of the defender's engagement capacity.

Additional benefits for the attacker include revealing the locations of air and missile defense systems as well as creating a negative psychological effect. Such attacks are most often carried out using inexpensive air threats. For example, the unit cost of a Shahed-136 unmanned aerial vehicle is on the order of tens of thousands of dollars [1], whereas a single surface-to-air interceptor missile currently costs hundreds of thousands or even several million dollars, as reported in the official U.S. Department of Defense budget documentation [2] and other open-source defense analyses [3–6].

The use of unmanned systems in saturation attacks is widespread. Evidence of this can be seen in the ongoing conflict in Ukraine. Analyses conducted by the Institute for Science and International Security (cf. Fig. 5.1) clearly demonstrate that the employment of UAS in this conflict is steadily increasing [7]. The key to countering saturation attacks is effective management of the available defensive resources. This problem is well known in the literature as the Weapon-Target Assignment (WTA) problem. [10]

5.2 THE WTA PROBLEM

In general terms, the WTA problem can be defined as the optimal allocation of a set of weapons to a set of targets (either for offensive or defensive purposes) in such a way that the expected survival value of the targets is minimized [8]. In the context of air and missile defense, the problem consists in assigning the available interceptors to incoming aerial threats in such a way as to maximize the probability of kill (Pk) for each individual weapon system or for the entire defensive formation, depending on the configuration.

A comprehensive formulation of the problem requires consideration of numerous aspects, including the number and types of weapons, the number of interceptors available, the ranges of both threats and interceptors, speed, time-to-intercept, and Pk and many other parameters [9].

The WTA problem is commonly classified into Static Weapon-Target Assignment (SWTA) and Dynamic Weapon-Target Assignment (DWTA) [8]. In SWTA, time is

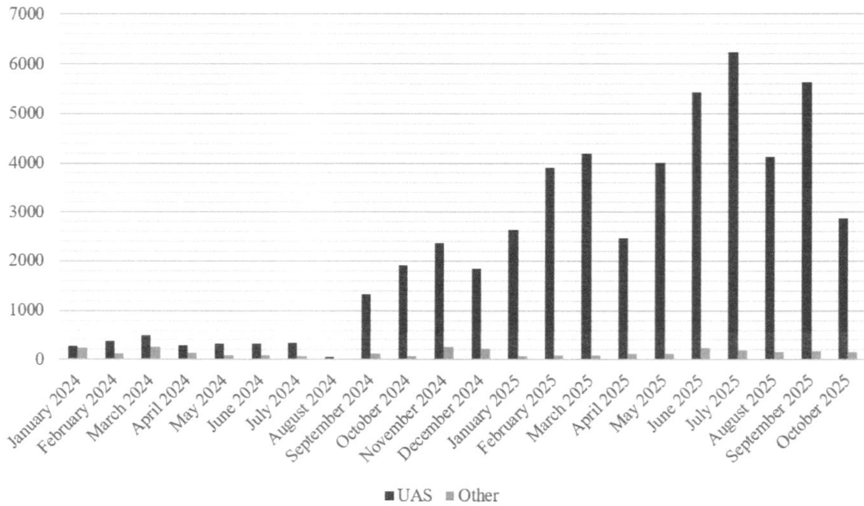

Figure 5.1 The use of unmanned aerial vehicles compared to other means in attacks on Ukrainian infrastructure facilities over.

not treated as a variable. The scenario is defined by a known number of weapons, known number of targets, and known probabilities of kill. The task is to maximize the effectiveness of defense at a specific instant, without considering future engagements. In contrast, DWTA extends SWTA by introducing time as a variable. As time evolves, positions of weapons and targets, their numbers, threat levels, and ammunition states all change, influencing the Pk [8]. While one might attempt to decompose a DWTA instance into a series of SWTA subproblems, the combination of solutions to separate stages will not, in general, yield the globally optimal solution for the entire operation [11]. DWTA can be further categorized into two main classes [9]:

1. m-stages—the problem is divided into multiple stages of SWTA, but future numbers and types of targets are assumed to follow a probability distribution. The objective is to allocate weapons in the first stage and reserve sufficient resources for subsequent stages to maximize overall effectiveness.
2. Shoot-Look-Shoot (SLS)—a multi-stage SWTA approach that explicitly considers the possibility of surviving targets ("leakers") and allows re-engagement in subsequent stages. Similarly to m-stages, the solution provides guidance on how to allocate and reserve weapons at each stage.

In summary, SWTA aims to maximize defensive effectiveness at a single instant, whereas DWTA seeks to optimize resource allocation globally over the entire defensive operation, taking into account opportunities for re-engagement and outcomes from previous stages.

Solutions to the WTA problem are generally divided into two categories: exact algorithms and heuristic/metaheuristic algorithms [8]. Exact methods are effective for small-scale problems, but as problem complexity grows, computational requirements increase dramatically, making such methods impractical in real-time applications. In contrast, heuristic and metaheuristic approaches are better suited for decision-support and command systems where computational time is critical. One of the earliest attempts to solve the WTA problem was made by Alan S. Manne in 1957 [12]. He proposed a nonlinear formulation aimed at minimizing the expected survivability of targets, defined as:

$$\min_{x_{ij}} \quad \sum_{j=1}^{n} a_j \prod_{i=1}^{m} (1 - p_{ij} x_{ij}) \tag{5.1}$$

$$\text{s.t.} \quad \sum_{j=1}^{n} x_{ij} = 1, \quad i = 1, \ldots, m \tag{5.2}$$

$$x_{ij} \geq 0 \quad \forall\, i, j \tag{5.3}$$

where $a_j > 0$ denotes the relative value or priority of target j, $p_{ij} \in [0,1]$ is the probability that weapon i successfully destroys target j, x_{ij} is a decision variable indicating the fraction of weapon i assigned to target j, subject to the constraint that each weapon must be fully assigned to one target.

The objective function minimizes the weighted sum of expected target survivabilities, where the product term in Eq. (5.1) represents the probability that target j survives attacks from all m weapons.

Under the simplifying assumptions that assignments are binary ($x_{ij} \in \{0,1\}$) and that all weapons have the same success probability ($p_{ij} = p$), the problem reduces to a discrete linear formulation:

$$\min_{y_j} \quad \sum_{j=1}^{n} a_j (1-p)^{y_j} \tag{5.4}$$

$$\text{s.t.} \quad \sum_{j=1}^{n} y_j = m \tag{5.5}$$

$$y_j \in \mathbb{Z}_{\geq 0} \tag{5.6}$$

where $y_j = \sum_{i=1}^{m} x_{ij}$ represents the total number of weapons assigned to target j. In this formulation, the term $(1-p)^{y_j}$ denotes the probability that target j survives y_j independent attack attempts.

Following this, WTA began to be extensively studied using exact methods such as enumerative algorithms, branch-and-bound, maximal marginal return, and dynamic programming [13]. While effective for static and well-defined problems, these approaches quickly became infeasible as problem sizes and complexity increased. In 1986, it was proven that WTA is NP-complete [14], which shifted research toward heuristic and metaheuristic methods aimed at finding approximate solutions. Considering the computational cost of exact methods and current computational capabilities, heuristic approaches are often the only practical option. Algorithms explored include neural networks, genetic algorithms, tabu search, particle swarm optimization, Very Large Scale Neighborhood (VLSN), among others [15, 16]. With the rise of artificial intelligence and machine learning, Reinforcement Learning (RL) in particular has become a promising approach due to its ability to adapt dynamically and continuously optimize performance [9].

5.3 SIMULATED ENVIRONMENT

In the presented work the missile is modeled as a generic short-range air-defense missile. The system is equipped with a two-stage rocket motor, providing propulsion during both the boost and sustain phases of flight. Control is provided by canard control surfaces, which enable rapid generation of aerodynamic moments and ensure high maneuverability. The flight dynamics are represented by a five-degree-of-freedom (5-DoF) model that accounts for translational and rotational motion in space while setting roll angle rate to zero (the roll angle is assumed to be actively stabilized). The missile guidance is organized into three stages—boost, midcourse, and terminal. During the midcourse phase, an optimal control law is employed, using information supplied by the ground radar network. In the terminal phase, guidance is performed using proportional navigation (PN), which uses the Radio Frequency (RF) seeker's measured radial velocity and line-of-sight angular rate as inputs. Selected characteristics of the proposed generic missile are shown in Fig. 5.2. Stabilization and guidance are implemented using a three-loop gain-scheduled autopilot [17]. This architecture ensures appropriate control dynamics across varying flight conditions, enabling the missile to remain stable and to effectively intercept targets over a wide range of operating parameters.

The guidance model of the proposed missile employs the equations of optimal guidance methods from [18], expressed by the relation:

$$\ddot{x}(t) = \frac{K_1}{t_g}\left[\dot{x}(t_f) - \dot{x}(t_0)\right] + \frac{K_2}{t_g^2}\left[x(t_f) - x(t_0) - \dot{x}(t_0)\,t_g\right] \tag{5.7}$$

$$\ddot{y}(t) = \frac{K_1}{t_g}\left[\dot{y}(t_f) - \dot{y}(t_0)\right] + \frac{K_2}{t_g^2}\left[y(t_f) - y(t_0) - \dot{y}(t_0)\,t_g\right] \tag{5.8}$$

$$\ddot{z}(t) = \frac{K_1}{t_g}\left[\dot{z}(t_f) - \dot{z}(t_0)\right] + \frac{K_2}{t_g^2}\left[z(t_f) - z(t_0) - \dot{z}(t_0)\,t_g\right] \tag{5.9}$$

where the gains are $K_1 = -2$ and $K_2 = 6$. An example trajectory of the generic missile guiding the linear target is shown in Fig. 5.3.

The dynamics of generic targets are modeled as a point-mass, first-order inertial system with a limited acceleration response. The guidance strategy is based on PN aimed at a defended asset (DA). To simulate deception, the targets perform random evasive maneuvers at unpredictable times during its flight path. Two different targets are assumed. The first is a generic UAS with a speed of 50 m/s and an acceleration saturation level of ±4 g, and the second is a generic CM with a speed of 300 m/s and an acceleration saturation level of ±6 g. An example tactical scenario with UAS and CM trajectories directed toward a DA is presented in Fig. 5.4. The DA is marked

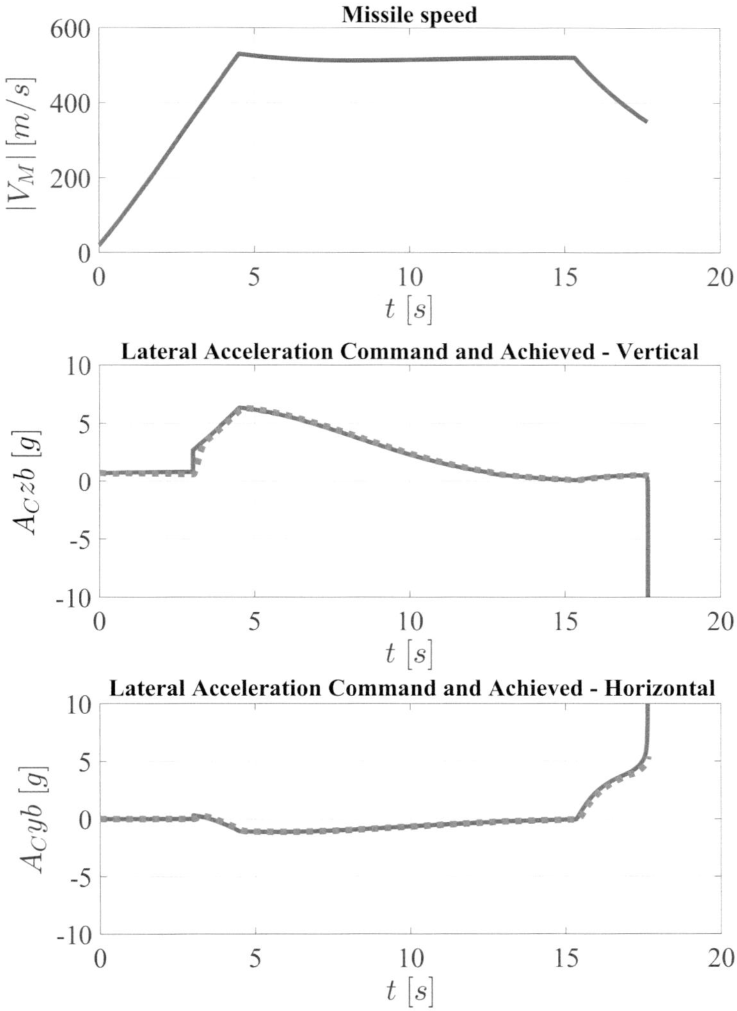

Figure 5.2 Selected dynamic characteristics of the proposed generic missile.

with a small circle in the middle, while the footprints of the launchers' engagement zones are marked with larger circles, respectively.

Following Guo et al. [19], the threat degree W_j is constructed from utility components that reflect closing range, target speed, and angular geometry, respectively. The original equation was extended with an additional logical variable A_{DA} which excludes targets that cannot reach the DA:

$$W_j = A_{DA}\left[\lambda_1 u_1^j(R_C) + \lambda_2 u_2^j(V_T) + \lambda_3 u_3^j(\delta)\right] \tag{5.10}$$

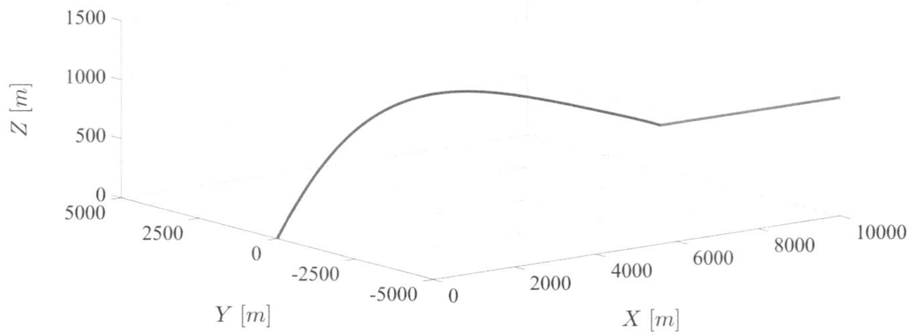

Figure 5.3 Example trajectory of the generic missile guiding the linear target.

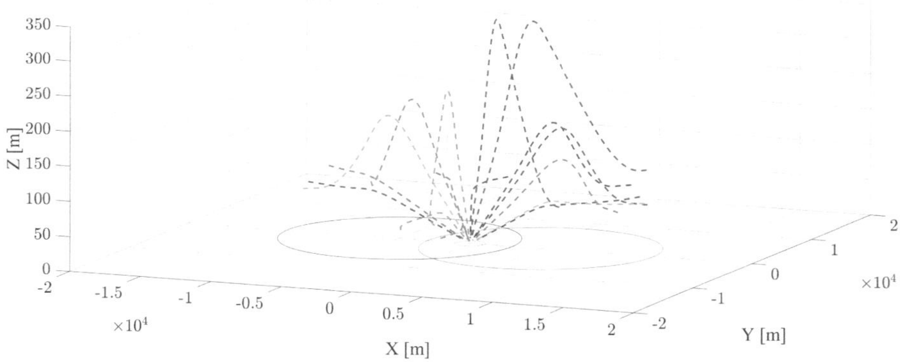

Figure 5.4 Example tactical scenario with UAS and CM trajectories directed toward a defended asset within the engagement zones of two launchers.

$$A_{DA} = \begin{cases} 1 & \text{if target} j \text{ will be able to achieve DA} \\ 0 & \text{if target} j \text{ will not be able to achieve DA} \end{cases} \tag{5.11}$$

The variable A_{DA} is evaluated in real time during the simulation through multiple runs of a virtual target guided toward the DA using the PN guidance law, while accounting for the target's maximum maneuvering capabilities (Fig. 5.5). Incorporating the variable enables a more efficient and balanced allocation of defensive resources (missiles) within an integrated defense system protecting multiple DAs. This approach accounts for the possibility that a given target may actually be heading toward a different DA covered by another launcher capable of engaging it. As a result, the inclusion of A_{DA} helps prevent situations in which the nearest launcher is

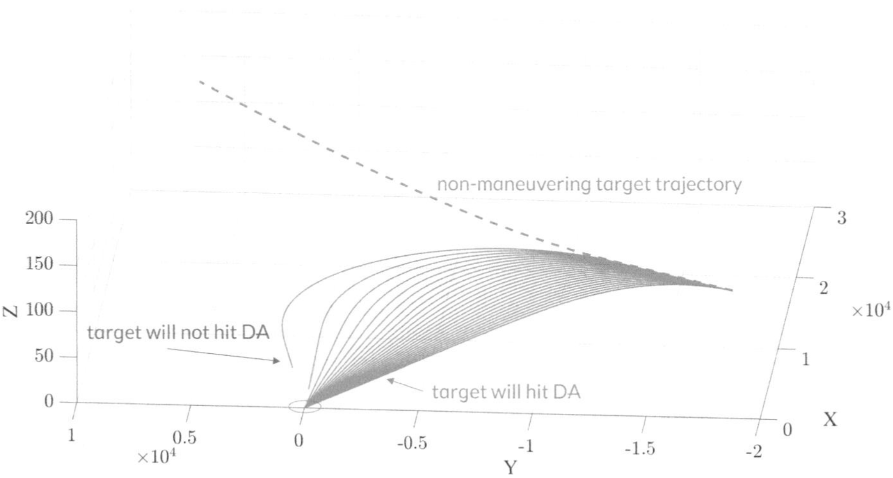

Figure 5.5 Assumptions underlying the logical mask A_{DA} for a single-target trajectory in the vicinity of a defended asset (DA).

disproportionately tasked with all engagements rapidly depleting its ammunition while other launchers remain idle and fully loaded.

$$u_1(R_C) = \begin{cases} 1 & \text{if } R_C \leq a_1 \\ \dfrac{b_1 - R_C}{b_1 - a_1} & \text{if } a_1 < R_C \leq b_1 \\ 0 & \text{if } R_C > b_1 \end{cases} \tag{5.12}$$

$$u_2(V_T) = \begin{cases} 0 & \text{if } V_T \leq a_2 \\ \dfrac{V_T^2 - a_2^2}{b_2^2 - a_2^2} & \text{if } a_2 < V_T \leq b_2 \\ 1 & \text{if } V_T > b_2 \end{cases} \tag{5.13}$$

$$u_3(\delta) = \cos(\delta) \tag{5.14}$$

Fig. 5.6 presents an illustrative overview of threat-degree computation and its implications. The left panel shows the trajectories of a CM, an UAS, and an unknown linear target with respect to a centrally located DA. The right column shows, from top to bottom, the instantaneous component terms $w_R = u_1(R_C)$, $w_V = u_2(V_T)$, and $w_\delta = u_3(\delta)$ together with the aggregate $W(t) = A_{DA}\left[\lambda_1 w_R + \lambda_2 w_V + \lambda_3 w_\delta\right]$ for each target type. The time courses of the components follow their parameterizations in u_1, u_2, and u_3. In the unknown linear case, the effect of the DA logical variable is visible because after the target passes the DA we have $A_{DA} = 0$ and the aggregate $W(t)$ becomes zero. The assumptions for the probability of kill calculations are shown in Fig. 5.7.

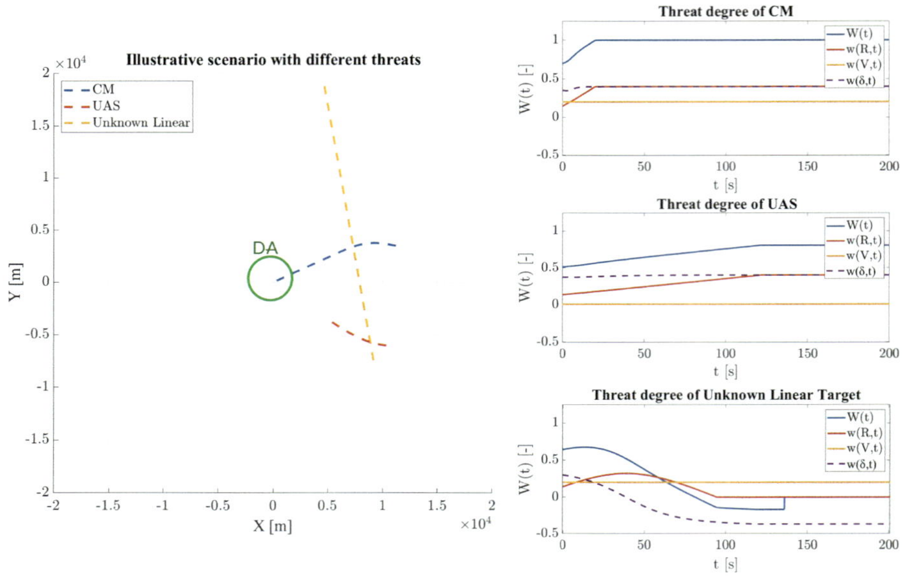

Figure 5.6 Example scenario illustrating the computation of the threat degree W_j.

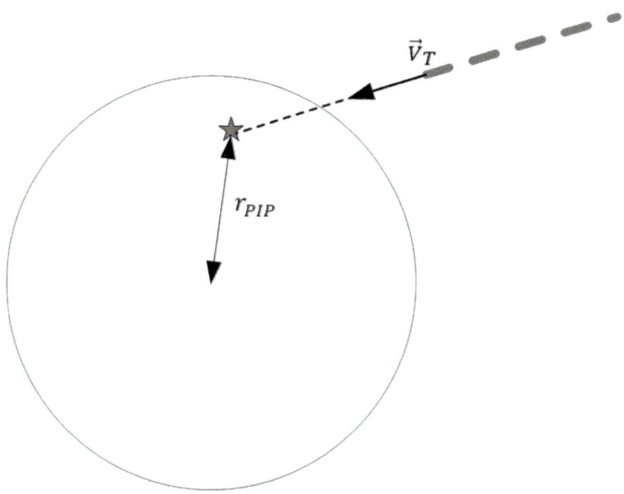

Figure 5.7 Probability of kill calculation assumptions.

$$i = (1, 2..., N_m) \qquad j = (1, 2, ..., N_t) \tag{5.15}$$

$$Pk_{ij}(t_{\text{launch}_i}) = \begin{cases} 0.95 & \text{if } r_{PIP_{ij}}(t_{\text{launch}_i}) \leq R_1(j) \\ 0 & \text{if } r_{PIP_{ij}}(t_{\text{launch}_i}) \geq R_2(j) \\ 0.95 \cdot e^{-\alpha \cdot \left(r_{PIP_{ij}}(t_{\text{launch}_i}) - R_1(j)\right)} & \text{otherwise} \end{cases} \tag{5.16}$$

$$t_{\text{launch}_i} = t + (i-1) \cdot t_{cd} \tag{5.17}$$

$$P_k = \begin{bmatrix} Pk_{11}\left(t_{\text{launch}_1}\right) & Pk_{12}\left(t_{\text{launch}_1}\right) & \cdots & Pk_{1j}\left(t_{\text{launch}_1}\right) \\ Pk_{21}\left(t_{\text{launch}_2}\right) & Pk_{22}\left(t_{\text{launch}_2}\right) & \cdots & Pk_{2j}\left(t_{\text{launch}_2}\right) \\ Pk_{31}\left(t_{\text{launch}_3}\right) & Pk_{32}\left(t_{\text{launch}_3}\right) & \cdots & Pk_{3j}\left(t_{\text{launch}_3}\right) \\ \vdots & \vdots & \ddots & \vdots \\ Pk_{i1}(t_{\text{launch}_i}) & Pk_{i2}(t_{\text{launch}_i}) & \cdots & Pk_{ij}(t_{\text{launch}_i}) \end{bmatrix}_{N_m} \tag{5.18}$$

where t_{launch_i} is the time of launch of i-th missile and t_{cd} is the time of launcher's cooldown.

For the purpose of the WTA benchmark, the tactical situation presented in Fig. 5.4 is assumed. Two launchers are positioned symmetrically, 5 km apart, relative to the DA. The adversary conducts a saturation attack involving eight unmanned aircraft systems (UAS) and eight CMs. Each launcher is equipped with eight missiles that can be fired sequentially, with a 3-second delay between launches due to cooldown requirements.

The scenario illustrated in the figure represents a fixed configuration of the launchers and the DA. However, the trajectories of the air threats shown constitute only a representative example. For the purpose of subsequent analyses, a broader set of attack instances with varying threat trajectories was generated to support comprehensive evaluation of the proposed methods.

5.4 PROBLEM FORMULATION AND SOLUTION APPROACHES

We address the DWTA problem—a time-expanded extension of the classical WTA introduced by Manne [12]. The DWTA seeks to allocate missile-target-time assignments to maximize overall engagement effectiveness under operational and physical constraints.

5.4.1 PROBLEM FORMULATION

Let

$$\mathscr{I} = \{1, \ldots, N_M\} \qquad \text{(set of missiles)} \tag{5.19}$$

$$\mathscr{J} = \{1, \ldots, N_T\} \qquad \text{(set of targets)} \tag{5.20}$$

$$\mathscr{T} = \{1, \ldots, N_H\} \qquad \text{(set of discrete time slots)} \tag{5.21}$$

The inputs are:

$$W \in \mathbb{R}_{+}^{N_T \times N_H} \quad P \in [0,1]^{N_M \times N_T \times N_H} \quad R \in \{0,1\}^{N_M \times N_T \times N_H} \quad \kappa \in \mathbb{Z}_{\geq 0}^{N_H} \tag{5.22}$$

where W_{jt} is the time-varying *threat degree* (priority) of target j at slot t, P_{ijt} is the probability of successful interception of j by missile i at t, R_{ijt} is the feasibility mask based on thresholds or reachability, and κ_t is the per-slot launch capacity.

The **assignment value** for a missile-target-time triple is defined as:

$$V_{ijt} = W_{jt} P_{ijt} \tag{5.23}$$

The decision variable x_{ijt} indicates assignment of missile i to target j at time t. The objective is:

$$F(X) = \sum_{t \in \mathscr{T}} \sum_{j \in \mathscr{J}} \sum_{i \in \mathscr{I}} V_{ijt} x_{ijt} \tag{5.24}$$

subject to:

$$\sum_{t \in \mathscr{T}} \sum_{j \in \mathscr{J}} x_{ijt} \leq 1 \quad \forall i \in \mathscr{I} \quad \text{(missile used at most once)} \tag{5.25}$$

$$\sum_{t \in \mathscr{T}} \sum_{i \in \mathscr{I}} x_{ijt} \leq 1 \quad \forall j \in \mathscr{J} \quad \text{(target engaged at most once)} \tag{5.26}$$

$$\sum_{i \in \mathscr{I}} \sum_{j \in \mathscr{J}} x_{ijt} \leq \kappa_t \quad \forall t \in \mathscr{T} \quad \text{(slot capacity)} \tag{5.27}$$

$$x_{ijt} \leq R_{ijt} \quad \forall i,j,t \quad \text{(feasibility)} \tag{5.28}$$

$$x_{ijt} \in \{0,1\} \quad \forall i,j,t \tag{5.29}$$

5.4.2 SOLUTION APPROACHES

We analyze three complementary approaches for solving DWTA: (a) Integer Linear Programming (ILP)—exact but computationally expensive; (b) Q-Learning—a tabular implementation of a classic RL method; and (c) Proximal Policy Optimization (PPO)—a policy-gradient method with feasibility-aware extensions. All operate on the same DWTA instances and input tensors (W, P, R, κ).

5.4.2.1 Integer Linear Programming (ILP)

ILP is a well-established exact optimization technique that guarantees globally optimal solutions for combinatorial assignment problems [20]. It is widely used as a baseline in WTA research due to its optimality guarantees and interpretability.

In this work, the ILP formulation solves Eq. (5.24) exactly, subject to constraints (5.25)–(5.29), using a standard integer programming solver integrated within the simulation framework. The coefficient vector is set to $c = -V(:)$ in order to convert the original maximization problem into the minimization form required by the solver. The ILP configuration and solver settings are summarized in Table 5.1.

Table 5.1
ILP Configuration for DWTA Experiments

Parameter	Value
Number of missiles N_M	4-64
Number of targets N_T	4-64
Number of time slots N_H	20
Slot capacity κ_t	2 (constant)
Thresholds τ_W, τ_P	0.0
Feasibility mask R_{ijt}	defined by thresholds
Solver	`intlinprog` (MATLAB R2024b)
Integrality tolerance	10^{-6}
RelativeGapTolerance	0
CutGeneration	`intermediate`
Heuristics	`advanced`

The ILP formulation guarantees an optimal solution for each DWTA instance and provides a reference solution for RL methods. However, the computational complexity grows rapidly with the problem size ($O(N_M N_T N_H)$), which can limit its applicability for large-scale or real-time scenarios.

5.4.2.2 Q-Learning (QL)

Q-learning [21] is one of the foundational model-free RL algorithms. It aims to learn an action-value function $Q(s,a)$ that estimates the expected cumulative reward obtained by taking action a in state s and subsequently following an optimal policy. A key property of Q-learning is that it does not require a model of the environment, relying instead on sampled interactions. Under mild conditions, tabular Q-learning converges to the optimal action-value function $Q^\star$ and thus to an optimal policy $\pi^\star$.

In its standard form, Q-learning updates the Q-values iteratively according to observed transitions, rewards, and a learning rate, while balancing exploration and exploitation through strategies such as ε-greedy action selection. The tabular variant is particularly suitable for discrete state-action spaces, as it maintains an explicit Q-table indexed by state-action pairs.

In this work, we adapt tabular Q-learning to solve the discrete, time-expanded variant of DWTA problem, formulated consistently with other presented approaches. The agent interacts with an environment that encodes the DWTA constraints directly, including missile availability, per-slot capacity κ_t, and admissibility mask R_{ijt}. This eliminates the need for the agent to learn these feasibility conditions implicitly and reduces the learning problem to sequential assignment decisions under combinatorial constraints. Each action corresponds to a feasible missile-target-time assignment triple (i, j, t), and the reward signal reflects the assignment value V_{ijt}. The hyperparameters of the Q-learning agent are provided in Table 5.2.

Table 5.2
Hyperparameters of the Q-Learning Agent

Parameter	Value
Learning rate α	0.4
Discount factor γ	0.0 (bandit formulation)
Episodes	1500
Exploration schedule	$\varepsilon_{\text{ep}} = \max(0.02,\ 0.9/\sqrt{\text{ep}})$
Initialization	$Q \equiv 0$
Missile order	Random permutation per episode
Action selection	ε-greedy over feasible set $\mathscr{F}_i(s)$

The learning process proceeds episodically: in each episode, missiles are processed in a random order, and for each missile i the algorithm constructs the feasible action set

$$\mathscr{F}_i(s) = \{(j,t) \in \mathscr{J} \times \mathscr{T} : u_t < \kappa_t,\ j \text{ unassigned},\ R_{ijt} = 1\} \tag{5.30}$$

where u_t denotes the number of missiles already assigned in slot t. The agent then selects an action from $\mathscr{F}_i(s)$ using an ε-greedy policy:

$$(j,t) = \begin{cases} \text{uniform sample from } \mathscr{F}_i(s) & \text{with probability } \varepsilon \\ \arg \max\limits_{(j',t') \in \mathscr{F}_i(s)} Q_{ij't'} & \text{with probability } 1-\varepsilon \end{cases} \tag{5.31}$$

Upon executing the action, the agent receives an immediate reward

$$r = V_{ijt}, \tag{5.32}$$

and the Q-value is updated according to the standard tabular rule (bandit variant with $\gamma = 0$):

$$Q_{ijt} \leftarrow Q_{ijt} + \alpha\big(r - Q_{ijt}\big) \tag{5.33}$$

The final solution returned by Q-learning is assessed using the objective function $F(X)$ defined in Eq. (5.24) computed for the assignment obtained at the end of the episode.

Training is performed over multiple episodes, with a decaying exploration rate ε_{ep}. and the learning rate α. The feasibility mask R_{ijt} is imported from the ILP formulation when available to ensure consistent problem instances across all methods.

While tabular Q-learning does not guarantee global optimality, it offers a lightweight, model-free alternative that can scale more efficiently to larger DWTA instances than ILP and serves as an interpretable baseline for evaluating policy-gradient methods such as PPO.

5.4.2.3 Proximal Policy Optimization (PPO)

PPO [22] belongs to the family of *policy gradient* methods in RL. In this paradigm, the agent learns a stochastic policy $\pi_\theta(a \mid s)$, parameterized by θ, that maps states s from the environment to a probability distribution over actions a. The goal is to maximize the expected cumulative reward by updating the policy parameters directly.

A key component of PPO is the probability ratio:

$$r_t(\theta) = \frac{\pi_\theta(a_t \mid s_t)}{\pi_{\theta_{\text{old}}}(a_t \mid s_t)} \tag{5.34}$$

where a_t is the action taken at time step t, s_t is the state at time t, $\pi_{\theta_{\text{old}}}$ is the policy before the update, and π_θ is the new policy being optimized.

The clipped surrogate objective used to stabilize training is defined as:

$$L_{\text{clip}}(\theta) = \mathbb{E}\left[\min\left(r_t(\theta)\hat{A}_t,\ \text{clip}(r_t(\theta), 1-\varepsilon_c, 1+\varepsilon_c)\hat{A}_t\right)\right] \tag{5.35}$$

where ε_c is the clipping coefficient and $\hat{A}_t$ is the advantage function estimating the relative value of action a_t in state s_t.

The advantage is computed using *generalized advantage estimation* (GAE):

$$\hat{A}_t = \sum_{\ell \geq 0} (\gamma\lambda)^\ell \, \delta_{t+\ell} \tag{5.36}$$

where $\gamma \in [0,1)$ is the discount factor $\lambda, \in [0,1]$ is the GAE parameter, and

$$\delta_t = r_t + \gamma V_\psi(s_{t+1}) - V_\psi(s_t) \tag{5.37}$$

is the temporal-difference error. Here $V_\psi(s)$ denotes the state-value function parameterized by ψ.

The full PPO objective combines the clipped surrogate loss, a value function loss, and an entropy regularization term:

$$\mathscr{L}(\theta,\psi) = \mathbb{E}\left[-L_{\text{clip}}(\theta)\right] + c_v\left(V_\psi(s_t) - \hat{R}_t\right)^2 - c_e\mathscr{H}\left(\pi_\theta(\cdot \mid s_t)\right) \tag{5.38}$$

where c_v and c_e are scalar weights, $\hat{R}_t$ is the empirical return, and $\mathscr{H}(\pi_\theta)$ denotes the entropy of the policy, which promotes exploration.

5.4.2.3.1 Heuristic Correction of Illegal Actions

During interaction with the DWTA environment, the policy π_θ may propose *illegal* actions, i.e., assignments that violate feasibility constraints (e.g., exceeding slot capacity, selecting an already assigned target, or violating the admissibility mask R_{ijt}). To address this, our PPO framework incorporates a heuristic correction mechanism that adaptively modifies invalid actions during the decision-making process.

For a given missile i in state s_t, let the feasible action set be defined as:

$$\mathscr{F}_i(s_t) = \{(j,t) \in \mathscr{J} \times \mathscr{T} : u_t < \kappa_t,\ j \text{ unassigned},\ R_{ijt} = 1\} \tag{5.39}$$

where u_t denotes the number of missiles already assigned in time slot t.

If the policy proposes an action $\tilde{a}_t = (\tilde{j}, \tilde{t})$ that belongs to the feasible set, it is executed as is. If not, the environment substitutes it with the best feasible action maximizing the expected assignment value:

$$a_t^\star = \arg \max_{(j,t) \in \mathscr{F}_i(s_t)} V_{ijt} \tag{5.40}$$

The reward signal is adjusted accordingly:

$$r_t = \begin{cases} \dfrac{V_{i\tilde{j}\tilde{t}}}{\overline{V}}, & \tilde{a}_t \in \mathscr{F}_i(s_t) \\ \dfrac{V_{ij^\star t^\star}}{\overline{V}} - \rho, & \tilde{a}_t \notin \mathscr{F}_i(s_t) \end{cases} \tag{5.41}$$

where $\rho \geq 0$ is a penalty applied when a correction occurs.

This heuristic correction serves two critical roles during training. First, it significantly improves the cumulative return F_{roll} by preventing the propagation of invalid actions and stabilizing policy updates. Second, it accelerates convergence and improves sample efficiency, particularly in large-scale DWTA instances with high-dimensional action spaces $(N_T N_H)$.

However, the use of heuristic correction also has implications for interpreting results. While it can improve performance even on previously unseen instances (W, P), part of this improvement may reflect the corrective support provided by the environment rather than the capabilities of the learned policy alone. Moreover, because the correction requires scanning the feasible action set $\mathscr{F}_i(s_t)$ at each step, it slightly increases per-action runtime, with more pronounced effects as problem size grows.

Importantly, the heuristic correction mechanism remains active during inference (policy deployment), not only during training. This design choice reflects a realistic operational scenario in which a supervisory layer ensures that WTAs remain feasible. As a result, the PPO agent's actions are always guaranteed to satisfy DWTA constraints, even if the policy itself proposes an infeasible assignment. The reported PPO results therefore reflect the combined effect of the learned policy and the environment's corrective logic, which together yield a more robust and operationally relevant decision-making process.

5.4.2.3.2 *Training Procedure*

The PPO agent was trained for 600 episodes, cf. Fig. 5.8. Training involved stochastic exploration (policy sampling), whereas evaluation used a deterministic policy derived from the trained actor. Training time ranged from several minutes to over half an hour depending on N_M, N_T, and N_H. The hyperparameters used in PPO training are provided in Table 5.3.

5.4.2.3.3 *Actor-Critic Structure and Hyperparameters*

The PPO agent uses a standard actor-critic architecture with a shared multilayer perceptron (MLP) backbone, following standard deep learning practice with Rectified

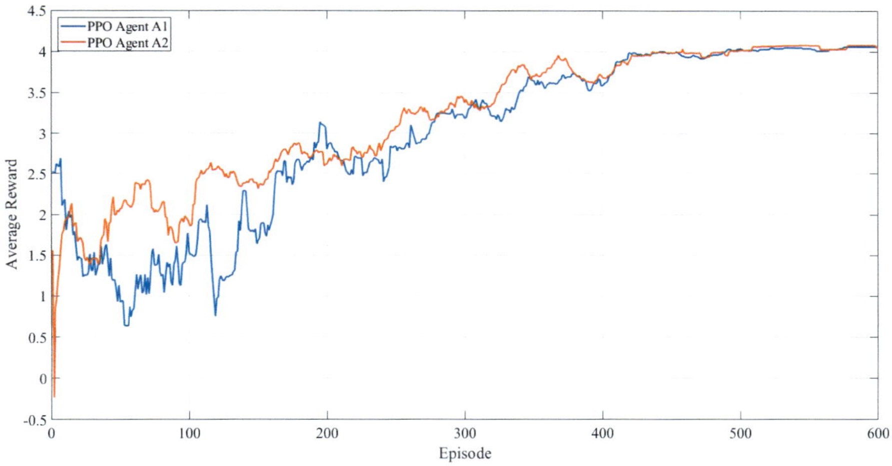

Figure 5.8 Average reward evolution for two PPO agents trained independently. Variance in early training reflects stochastic exploration, while convergence shows stabilization of policy performance.

Table 5.3
Hyperparameters Used in PPO Training

Parameter	**Value**
Training episodes	600
Discount factor γ	0.99
GAE parameter λ	0.95
Clipping factor ε_c	0.2
Entropy coefficient c_e	0.01
Value loss coefficient c_v	0.5
Batch size	256
Epochs per update	10
Actor/Critic LR	1×10^{-4}
Policy (eval)	Deterministic
Illegal action correction	Enabled

Linear Unit (ReLU) activations [23]. This backbone processes the state vector and extracts a compact latent representation that is then fed into two distinct outputs: the *actor* and the *critic* networks (cf. Table 5.4.).

The **actor network** parameterized by θ maps the current state s_t to a categorical probability distribution $\pi_\theta(a_t \mid s_t)$ over the discrete action space $\mathcal{J} \times \mathcal{T}$ where each action corresponds to a missile-target-time assignment. Sampling from this distribution allows the agent to explore different feasible assignments during training. During evaluation, the learned policy is applied deterministically by selecting the

Table 5.4
Neural Network Architecture of the Actor and Critic

Component	Specification
Input (state vector)	$N_T N_H + N_T N_H + N_T + N_H + N_M + 1 + N_T N_H$
Actor (policy network)	FC(256)-ReLU-FC(256)-ReLU-FC($N_T N_H$) > Softmax
Critic (value network)	FC(256)-ReLU-FC(256)-ReLU-FC(1)
Action distribution	Categorical over $\mathscr{J} \times \mathscr{T}$

action with the highest probability.

The **critic network**, parameterized by ψ, estimates the state-value function $V_\psi(s_t)$, which represents the expected return from state s_t under the current policy. This value estimate is used to compute the temporal-difference error and the advantage function $\hat{A}_t$ in GAE, stabilizing the gradient updates and improving sample efficiency.

The actor and critic share the same feature-extraction layers but maintain separate fully connected output layers. Both networks use ReLU activations and are trained jointly with the PPO objective.

The PPO agent operates on a structured but reduced state representation compared to the full decision space of size $N_M \times N_T \times N_H$ considered in the ILP formulation. Rather than encoding all possible missile-target-time assignments at once, the observation provided to the agent focuses on the currently processed missile, the values and feasibility of its potential engagements, the global assignment status, and the remaining slot capacities. This sequential formulation is a standard design choice in RL for large-scale combinatorial problems, as it allows the agent to operate within a tractable observation space while preserving all operational constraints of the DWTA problem. Consequently, although PPO does not solve the assignment globally in a single optimization step, it remains comparable to ILP and tabular Q-learning when evaluated using the same objective function $F(X)$ and feasibility conditions.

5.5 METHODS AND EVALUATION

For each scenario $s \in \mathscr{S}$, we consider three solvers $\mathscr{M} = \{\text{ILP}, \text{QL}, \text{PPO}\}$. Each method $m \in \mathscr{M}$ returns an assignment $X^{m,s}$ and a wall-clock time $T^{m,s}$.

Let $X^{\star,s}$ denote the optimal assignment (when available), provided by the ILP solver. For each method $m \in \mathscr{M}$ and scenario $s \in \mathscr{S}$, we evaluate its objective value $F\big(X^{m,s}\big)$. The optimal objective in scenario s is defined as:

$$F^\star(s) = F\big(X^{\star,s}\big) \tag{5.42}$$

We assume a quality tolerance $\varepsilon \in [0,1)$, which bounds the acceptable normalized loss, and a time budget $\tau > 0$, which limits the wall-clock runtime. We then define the admissible (feasible) set as:

$$\mathscr{A}^{\varepsilon,\tau}(s) = \{m \in \mathscr{M} \;:\; T^{m,s} \leq \tau,\; \delta^{m,s} \leq \varepsilon\} \tag{5.43}$$

The normalized quality loss $\delta^{m,s}$ is defined as:

$$\delta^{m,s} = \frac{F\big(X^{\star,s}\big) - F\big(X^{m,s}\big)}{F\big(X^{\star,s}\big)} \in [0,1] \tag{5.44}$$

Accordingly, the comparison of method performance is bi-criteria (quality-time), and we resolve it lexicographically in two complementary ways, evaluated over the admissible set $\mathscr{A}^{\varepsilon,\tau}(s)$: a *quality-first* rule and a *time-first* rule.

In the *quality-first* rule, given the admissible set $\mathscr{A}^{\varepsilon,\tau}(s)$, we select the method that minimizes the normalized quality loss $\delta^{m,s}$. Ties on δ are broken by choosing the one with the shorter runtime $T^{m,s}$. If $\mathscr{A}^{\varepsilon,\tau}(s) = \varnothing$, we assign the label NONE (no feasible method). This can be formalized as:

$$m_{\delta}^{\varepsilon,\tau}(s) \in \begin{cases} \arg\min\limits_{m\in\mathscr{A}^{\varepsilon,\tau}(s)} \big(\delta^{m,s}, T^{m,s}\big) & \mathscr{A}^{\varepsilon,\tau}(s) \neq \varnothing \\ \text{NONE} & \mathscr{A}^{\varepsilon,\tau}(s) = \varnothing \end{cases} \tag{5.45}$$

In the *time-first* rule, given the admissible set $\mathscr{A}^{\varepsilon,\tau}(s)$, we select the method that minimizes the runtime $T^{m,s}$. Ties on T are broken by choosing the one with the smaller normalized quality loss $\delta^{m,s}$:

$$m_{T}^{\varepsilon,\tau}(s) \in \begin{cases} \arg\min\limits_{m\in\mathscr{A}^{\varepsilon,\tau}(s)} \big(T^{m,s}, \delta^{m,s}\big) & \mathscr{A}^{\varepsilon,\tau}(s) \neq \varnothing \\ \text{NONE} & \mathscr{A}^{\varepsilon,\tau}(s) = \varnothing \end{cases} \tag{5.46}$$

For the scenario set $\mathscr{S}$, we report aggregates:

$$\bar{F}^{m} = \frac{1}{|\mathscr{S}|}\sum_{s\in\mathscr{S}} F\big(X^{m,s}\big) \qquad \bar{T}^{m} = \frac{1}{|\mathscr{S}|}\sum_{s\in\mathscr{S}} T^{m,s} \tag{5.47}$$

We define the *unconditional shares* (normalized by $|\mathscr{S}|$), which include the NONE category, under both the quality-first and time-first approaches, with the set of methods expanded to $m \in \mathscr{M} \cup \text{NONE}$, as follows:

$$\varpi_{\delta}^{m} = \frac{1}{|\mathscr{S}|}\sum_{s\in\mathscr{S}} \mathbf{1}\big[m_{\delta}^{\varepsilon,\tau}(s) = m\big] \tag{5.48}$$

$$\varpi_{T}^{m} = \frac{1}{|\mathscr{S}|}\sum_{s\in\mathscr{S}} \mathbf{1}\big[m_{T}^{\varepsilon,\tau}(s) = m\big] \tag{5.49}$$

Additionally, we report the unconditional shares corresponding to the **NONE** outcome, i.e., the fraction of scenarios where the admissible set is empty:

$$\varpi_{\delta}^{\text{NONE}} = \frac{1}{|\mathscr{S}|}\sum_{s\in\mathscr{S}} \mathbf{1}[\mathscr{A}^{\varepsilon,\tau}(s) = \varnothing] \tag{5.50}$$

$$\varpi_{T}^{\text{NONE}} = \frac{1}{|\mathscr{S}|}\sum_{s\in\mathscr{S}} \mathbf{1}[\mathscr{A}^{\varepsilon,\tau}(s) = \varnothing] \tag{5.51}$$

5.6 RESULTS

We evaluate the three solvers included in $\mathscr{M}$ over the shared scenario set $\mathscr{S}$ described earlier. The comparison follows the bi-criteria protocol introduced in Section 5.4: (a) solution quality measured by the achieved objective $F(X^{m,s})$ relative to the ILP optimum $F^{\star}(s)$ when available and (b) computational runtime $T^{m,s}$. Outcomes are summarized at two levels. First, aggregate runtimes and average objective values are reported to illustrate the baseline trade-off between optimality and computational speed. Second, the selector rules with admissibility thresholds (ε, τ) are applied to quantify, under operational constraints, which methods are preferred across scenarios.

5.6.1 COMPUTATION TIME

Figure 5.9 shows the computational runtime of the ILP solver and both RL methods across the scenario set $\mathscr{S}$. The results highlight clear scalability differences between the approaches. As problem size increases, the ILP runtime grows rapidly due to its combinatorial complexity, whereas both RL methods maintain runtimes within ranges compatible with near-real-time decision-making. PPO exhibits the shortest inference times overall, despite incorporating heuristic correction, which underscores its suitability for time-critical applications.

For the Q-learning method, training time is included in the overall runtime because the learned policy does not generalize across the scenario set with varying threat priorities and kill probabilities. This significantly affects the observed performance characteristics of the method.

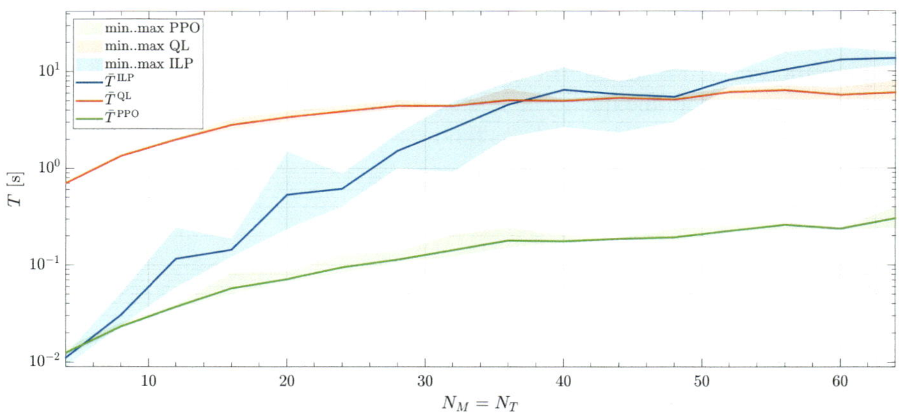

Figure 5.9 Computational runtime of ILP and two RL methods across the scenario set $\mathscr{S}$.

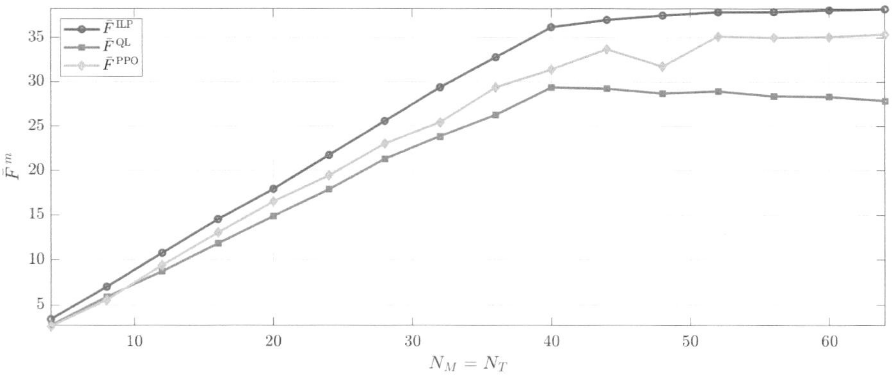

Figure 5.10 Average reward evolution for Q-learning and PPO agents during training.

5.6.2 SOLUTION QUALITY

The quality of the obtained assignments is evaluated using the objective $F(X^{m,s})$ defined in Eq. (5.24). Figure 5.10 presents the average reward obtained during the training process for both RL agents. PPO demonstrates faster convergence and stabilizes at higher reward levels than tabular Q-learning, narrowing the quality gap relative to the ILP baseline, especially for medium-sized problem instances. This indicates that policy-gradient methods with feasibility-aware corrections can learn effective assignment strategies in complex DWTA environments.

5.6.3 SELECTOR-BASED EVALUATION

To further analyze solver performance under operational constraints, selector rules defined in Section 5.5 are applied. These rules incorporate admissibility conditions based on maximum runtime τ and acceptable normalized quality loss ε. The analysis reports unconditional selection shares ϖ^m_δ and ϖ^m_T from Eq. (5.49), as well as the NONE shares from Eq. (5.51), which indicate cases where no solver meets both criteria simultaneously.

Figures 5.11 and 5.12 present selector outcomes for the *quality-first* rule. Results are shown for runtime thresholds $\tau = 1\,\text{s}$ and $\tau = 5\,\text{s}$ across quality tolerances $\varepsilon \in \{0.05, 0.1, 0.15, 0.2\}$. The vertical axis shows the normalized number of selections of each method within the scenario set, revealing how solver preference changes with problem scale and admissibility requirements. Gray-coded NONE segments denote scenarios where no solver satisfies $\mathscr{A}^{\varepsilon,\tau}(s) \neq \varnothing$.

Figures 5.13 and 5.14 show analogous results under the *time-first* rule. These highlight the shift in solver preference when minimizing runtime becomes the primary criterion. PPO emerges as the most frequently selected method under tight time budgets, while ILP gains dominance when the time constraint is relaxed and quality requirements are stringent. The NONE shares again capture scenarios where no solver meets the simultaneous time-quality conditions.

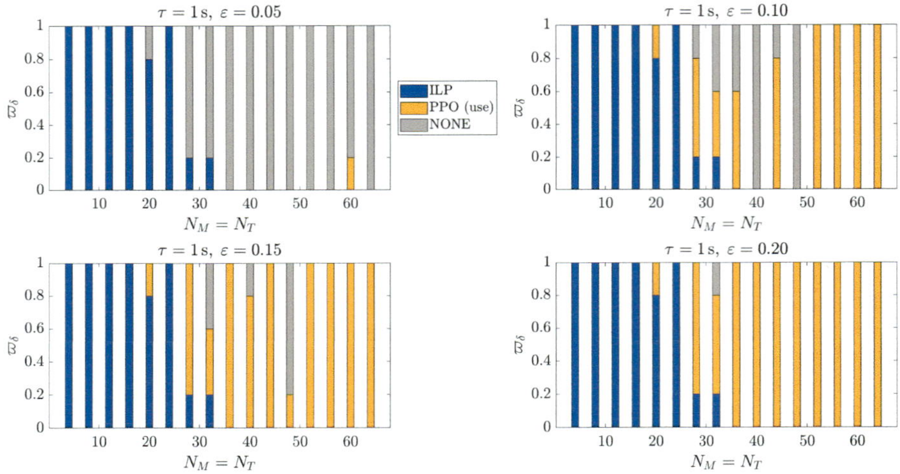

Figure 5.11 Selector outcomes—*unconditional shares* under the quality-first rule with $\tau = 1\,\mathrm{s}$ and various quality tolerances ε.

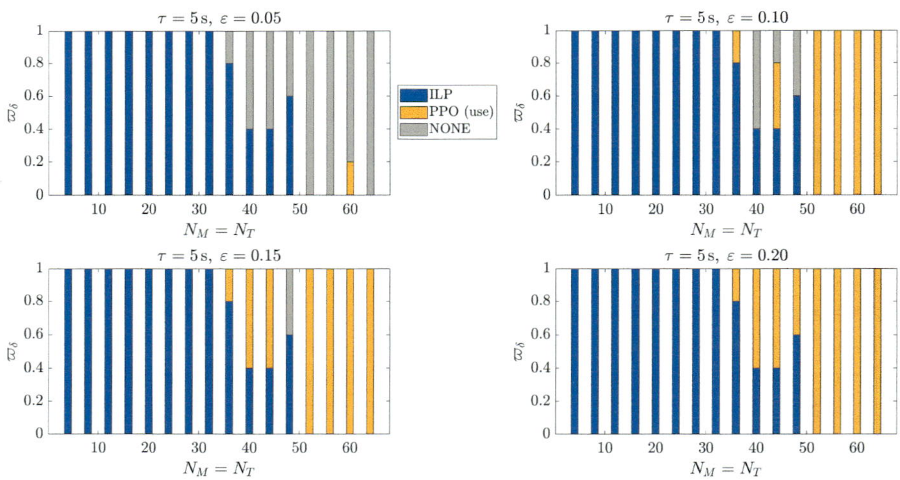

Figure 5.12 Selector outcomes—*unconditional shares* under the quality-first rule with $\tau = 5\,\mathrm{s}$ and various quality tolerances ε.

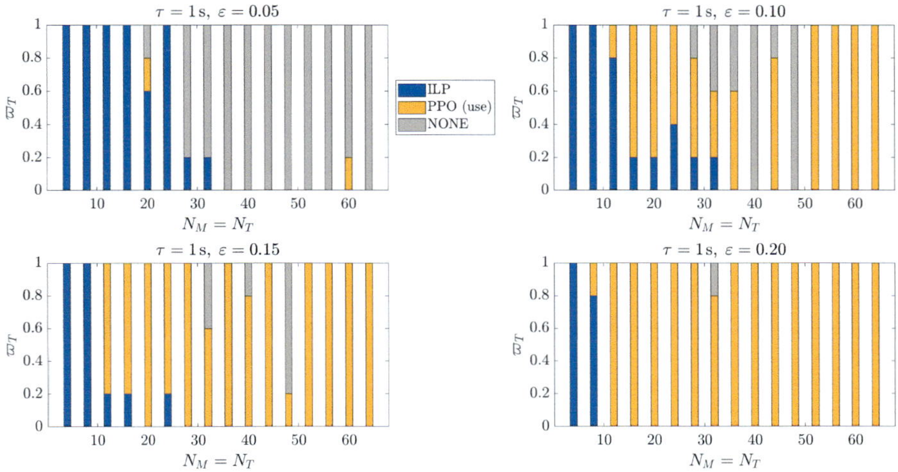

Figure 5.13 Selector outcome—*unconditional shares* under the time-first rule with $\tau = 1\,\mathrm{s}$ and various quality tolerances ε.

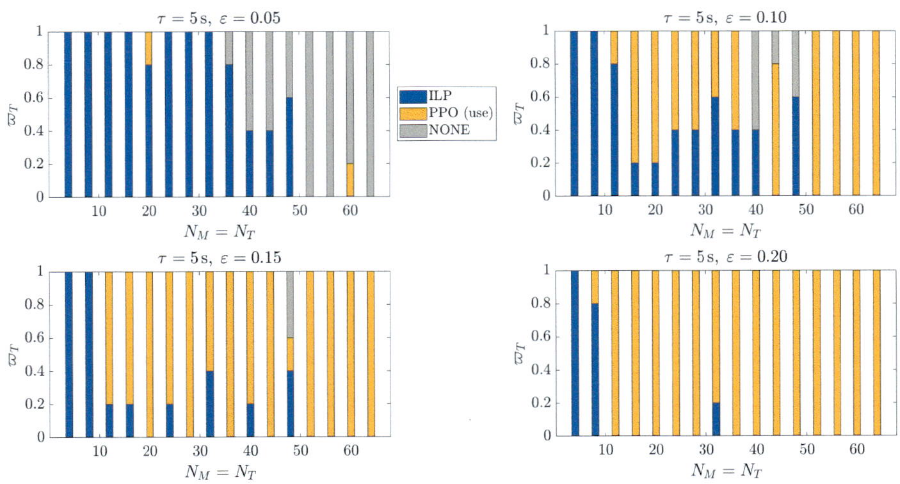

Figure 5.14 Selector outcomes—*unconditional shares* under the time-first rule with $\tau = 5\,\mathrm{s}$ and various quality tolerances ε.

5.6.4 TACTICAL SCENARIO VERIFICATION

In addition to aggregate metrics and selector outcomes, scenario effects can be verified directly at the trajectory level. A representative example is shown in Fig. 5.15. The top panel shows one missed target (one leaker) resulting from the firing sequence determined by the solution to the WTA problem, while the bottom panel shows two missed targets (two leakers) for the same scenario. The composite figure uses the same launcher-DA geometry, threat kinematics, and per-slot capacity settings for both panels. Only the missile-target-time assignments and launch timing differ, as determined by the selected method and admissibility constraints. The trajectories of targets and interceptors around the DA make intercepted tracks and leakers (missed targets) visible in state space. This qualitative view complements the quantitative

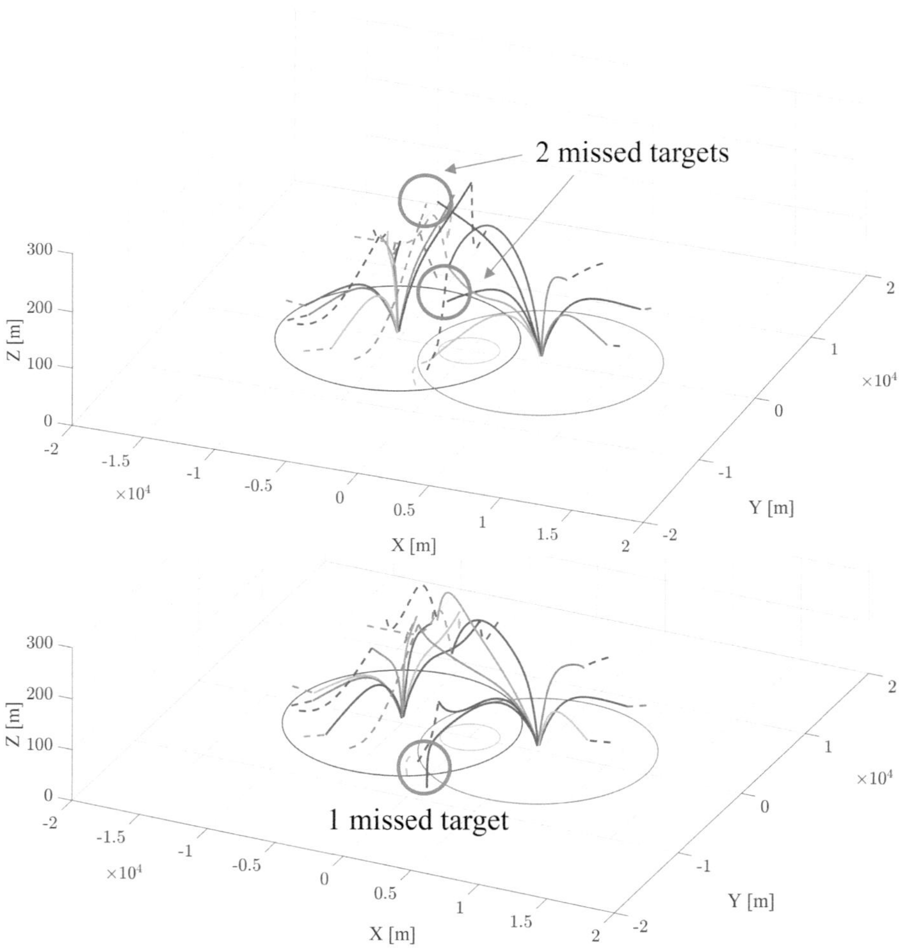

Figure 5.15 Trajectory-level verification in the same tactical scenario.

evaluation based on $F(X)$, W, P, and R and is consistent with the DA logical variable A_{DA}, which sets $W(t)$ to zero once a target can no longer reach the DA.

5.7 CONCLUSIONS

This study addressed robust resource allocation for counter-UAS defense under saturation attacks through a DWTA formulation and a physics-based simulation environment. In line with the article's title, the focus was on operationally useful allocation rules rather than purely theoretical optima. The comparative analysis of an exact ILP solver and two RL baselines (tabular Q-learning and a feasibility-aware PPO) clarified the core trade-off between solution quality and computational time under explicit admissibility thresholds.

A key contribution beyond the solvers themselves is the simulation framework, which enables consistent, repeatable verification of WTA solutions under dynamic threat kinematics and system constraints. By exposing time-varying threat priorities W and interception probabilities P as first-class tensors tied to feasibility and launch capacity, the environment supports end-to-end evaluation of both assignment logic and its inputs. In particular, it allows systematic stress-testing of alternative constructions of W (e.g., variants that emphasize proximity, velocity, intent, or DA geometry) and data-driven or model-based estimates of P, including range-dependent lethality curves and reachability masks.

The selector analysis grounded in admissibility (ε, τ) offers a practical lens for method choice under real decision timelines. It formalizes when exact optimization is preferable and when policy-based decision-making is adequate given time budgets, thereby connecting algorithmic performance to operational constraints. This framework can be readily extended to additional solvers and to multi-objective settings that include, e.g., interceptor expenditure or risk to DAs.

The RL components developed here should be viewed as baselines. The tabular Q-learning agent provides a simple, interpretable reference; the PPO policy, augmented with feasibility-aware correction, demonstrates how a supervised layer can stabilize training and deployment. Building on this foundation, future work can explore richer state encodings (e.g., graph representations of missile-target-time lattices), constrained or safe RL with primal-dual penalties, offline RL from ILP-generated datasets, and hierarchical policies that first select time slots and then target assignments. On the modeling side, integrating sensor fusion latency, adversarial deception models, and stochastic arrival processes would further align evaluation with operational realities.

Overall, the combination of an explicit DWTA formulation, a physically grounded simulation environment, and an admissibility-aware evaluation protocol provides a coherent basis for studying robust resource allocation in C-UAS defense. It supports not only fair comparisons between exact and learning-based methods but also principled iteration on the underlying components of the WTA problem—most notably the definitions of W and P—which ultimately drive tactical relevance.

AUTHOR CONTRIBUTIONS

Conceptualization, P.T. and M.K.; methodology, P.T.; software, M.K. and P.T.; validation, P.T.; formal analysis, P.T.; investigation, J.M.; resources, J.M. and M.K.; data curation, M.K. and P.T.; writing—original draft preparation, P.T., M.K., and J.M.; writing—review and editing, P.T.; visualization, M.K.; supervision, P.T.; project administration, P.T.; funding acquisition, P.T. All authors have read and agreed to the published version of the manuscript.

DATA AVAILABILITY STATEMENT

The data that support the findings of this chapter are available from the corresponding author, P.T., upon reasonable request.

FUNDING

This work was financed by the Military University of Technology under the research project UGB-531-0000-46-W200-22.

REFERENCES

1. Suicide drones are much cheaper to launch than to shoot down. 2023. Business Insider. https://www.businessinsider.com/suicide-drones-much-cheaper-launch-than-shoot-down-ukraine-nyt-2023-1?IR=T (accessed April 01, 2025).
2. Program Acquisition Cost by Weapon System, Fiscal Year 2024 Budget Request. March 2023. U.S. Department of Defense, Office of the Under Secretary of Defense. https://comptroller.war.gov/Portals/45/Documents/defbudget/FY2024/FY2024_Weapons.pdf (accessed October 30, 2025).
3. US greenlights precision rocket sale to Saudi Arabia amid rising drone threats in the Gulf. 2023. Defence Security Asia. https://defencesecurityasia.com/en/us-greenlights-precision-rocket-sale-to-saudi-arabia-amid-rising-drone-threats-in-the-gulf/ (accessed April 01, 2025).
4. AIM-120 AMRAAM Fact Sheet. 2023. U.S. Air Force. https://www.armyrecognition.com/military-products/army/missiles/tactical-missiles/amraam-aim-120-air-to-air-missile (accessed April 01, 2025).
5. AMRAAM AIM-120 Air-to-Air Missile. 2023. Army Recognition. https://armyrecognition.com/military-products/army/missiles/tactical-missiles/amraam-aim-120-air-to-air-missile (accessed April 01, 2025).
6. Poland inks a massive purchase of Piorun MANPADS. 2025. Defence24. https://defence24.com/industry/poland-inks-a-massive-purchase-of-piorun-manpads (accessed April 01, 2025).
7. Updated Analysis of Russian Shahed-136 Deployment Against Ukraine. 2025. Institute for Science and International Security. https://isis-online.org/uploads/isis-reports/documents/Updated_Analysis_of_Russian_Shahed_136_Deployment_Against_Ukraine_March_2025_Final.pdf (accessed April 01, 2025).
8. Andersen, A.C., K. Pavlikov, and T.A.M. Toffolo. 2022. Weapon-target assignment problem: Exact and approximate solution algorithms. *Annals of Operations Research* 312(2):581–606.

9. Li, J., G. Wu, and L. Wang. 2024. A comprehensive survey of the weapon target assignment problem: Model, algorithm, and application. *Engineering Applications of Artificial Intelligence* 137, part B.
10. Kline, A., D. Ahner, and R. Hill. 2019. The Weapon-Target Assignment problem. *Computers and Operations Research* 105(C):226–236.
11. Chen, J., B. Xin, Z.H. Peng. 2009. Evolutionary decision-making for the dynamic weapon–target assignment problem. *Science in China Series F: Information Sciences* 52(11):2006–2018.
12. Manne, A.S. 1957. A target-assignment problem. *Operations Research* 5(3):346–351.
13. Huaiping, C., L. Jingxu, C. Yingwu, et al. 2006. Survey of the research on dynamic weapon–target assignment problem. *Journal of Systems Engineering and Electronics* 17(3):559–565.
14. Lloyd, S.P., and H.S. Witsenhausen. Weapons allocation is NP-complete. In: Proc. of the 1986 Summer Computer Simulation Conference. Reno (NV, USA), 1986, pp. 1054–1058.
15. Kline, A., D. Ahner, and B. Lunday. 2020. A heuristic and metaheuristic approach to the static weapon-target assignment problem. *Journal of Global Optimization* 78:791–812.
16. Ahuja, R., A. Kumar, K. Jha, et al. 2003. Exact and Heuristic Methods for the Weapon-Target Assignment Problem. *MIT Sloan School of Management*, Working Paper 4464-03.
17. Bryson, J.T., and B.C. Gruenwald. 2022. Automated Gain-Scheduling Approach for Three-Loop Autopilot. U.S. Army Combat Capabilities Development Command Army Research Laboratory. https://apps.dtic.mil/sti/citations/trecms/AD1181622 (accessed May 07, 2025).
18. Lin, C.F. 1991. *Modern Navigation, Guidance, and Control Processing*. New York: Prentice Hall.
19. Guo, J., G. Hu, Z. Guo, et al. 2022. Evaluation model, intelligent assignment, and cooperative interception in multimissile and multitarget engagement. *IEEE Transactions on Aerospace and Electronic Systems* 58(4):3104–3115.
20. Wolsey, L.A. 1998. *Integer Programming*. New York: John Wiley & Sons.
21. Watkins, C.J., and P. Dayan. 1992. Q-learning. *Machine Learning* 8(3–4):279–292.
22. Schulman, J., F. Wolski, P. Dhariwal, et al. 2017. Proximal Policy Optimization Algorithms. https://arxiv.org/pdf/1707.06347 (accessed October 30, 2025).
23. Géron, A. 2018. *Uczenie maszynowe z użyciem Scikit-Learn i TensorFlow*. Gliwice: Helion (in Polish).

Section II

Aerial Target Detection, Recognition, and Identification Issues

6 Assessments on Drone and Hypersonic Weapons Surveillance Challenges and Their Solutions

Istvan Balajti

László Szilvássy
Doctoral School of Military Engineering, Ludovika University of Public Service, Budapest, Hungary

CONTENTS

6.1 Background 74
6.1.1 Current Challenges and Related Research Questions 74
6.1.2 Topic-Related Hungarian Historical Roots for Education 76
6.2 The Key Non-Controlled Challenges of Radar Designers 77
6.2.1 Challenges of Target Detection 78
6.2.2 Challenges of Target Track Initialization and Track Maintenance 80
6.2.3 Challenges of I5.0-Based AI-Supported IADS Applications 82
6.3 Drone Defense Characterization 82
6.4 Hypersonic Weapons Characterization 84
6.4.1 The Key Performances of the Hypersonic Weapons 85
6.4.2 The Analysis of Traditional Sensor/Radar Performances for HyW Detection 87
6.4.3 The Analysis of Radar Network Topology Suitable against HyW 88
6.4.4 Quasi-Monostatic Surveillance Radar Systems Based on the I5.0 Production 89
6.4.5 Q-M-Based VHF Surveillance Radar Networks for University-Level Education Propose 90
6.4.6 Q-M VHF Surveillance Radar Networks Augmented by Passive Radar 91
6.4.7 Multi-static VHF Surveillance Radars with Exploration of Ionosphere Reflection 93

DOI: 10.1201/9781003739234-8

6.4.8 Realization of LOS Q-M CubeSat-Type Surveillance Sensor/Radar Systems.......94
6.4.9 Near-Future IADS Sensors Vertical Realization Vision.......96
6.5 Conclusions.......98

6.1 BACKGROUND

Air defense, which had previously been accorded a low priority in the low-threat air campaigns that dominated before the war in Ukraine, has once again become a central tenet in defense planning [1]. The advent of novel military targets, such as “dim” targets, UAV (Unmanned Aerial Vehicles—later called drones), missiles, stealth technology, and hypersonic weapons within the air and space domains has led to an escalation in the complexity of the Integrated Air Defense System (IADS). Drones present a problematic target due to their very low radar cross-section (RCS), stealth operational capabilities, high range of maneuverability from zero to 10 G, and mass-produced types and implementations that challenge their recognition. Hypersonic defense systems are required to operate at speeds that necessitate rapid reaction and accurate response in the multi-domain environment beyond the operational characteristics of traditional sensor systems. The war in Ukraine has demonstrated the necessity for military analysts, interested civilians, and planners to recognize and adopt a thorough understanding of the emerging complex Industry 5.0 (I5.0)-based IADS concepts, to effectively communicate these threats to decision-makers at the tactical, operational, and strategic levels. Consequently, it is imperative to facilitate military operations and campaigns by integrating advanced multi-domain sensor network technologies on the battlefield. The advent of I5.0 has given rise to a plethora of engineering challenges, necessitating the development of innovative training programs for military engineers and commanders.

6.1.1 CURRENT CHALLENGES AND RELATED RESEARCH QUESTIONS

The most important challenge for solving this task is the training of commanders and military engineers to survive and be efficient on the battlefields of the 21st century. According to the authors, this paradigm can be managed efficiently by the synergy of mechatronics, exploiting the offered possibilities of Education 5.0 (E5.0) supported by artificial intelligence (AI). The technical challenges of radar engineers and scientists of our time can be grouped around two further central research questions. As humorously expressed, these are how to overcome the facts that: “the speed of light is not fast enough” and “the Earth is not flat”.

The solutions are related to research and development that offer reliable solutions for software-defined wireless sensor networks using machine/deep learning (ML/DL) techniques. While the development of sensing technologies and devices is triggered by inventions and demands, there is a growing need for high-performance sensing technologies for several specific dual-use applications of I5.0s. For instance, a recently emerged demand is for advanced 3D sensing, environmental monitoring, and supporting enhanced situational awareness [2]. The literature [3, 4] informs about dedicated algorithms that are underpinned by the theory of interferometry, microwave holography, polarimetry, and quantum sensing.

The countermeasures, from a theoretical standpoint by the classic radar equation, comprise numerous non-linear terms, necessitating real-time estimation of optimal signal-to-noise ratio operation, which must be multi-layered, encompassing all phases of flight detection, tracking, recognition, and interception. It requires an extension for the radar measurement degree of freedom (RMDF). The RMDF is a key requirement for surveillance volume extensions such as range, azimuth and elevation, Doppler frequency, wavelength/frequency band (multi-sensor fusion), polarization and real-time signal processing.

At present, air surveillance systems (ASS) assets such as radars, very low frequency detectors like acoustics, ultrasonic vs. optical and infrared sensors, communication equipment, and command centers are platform-based, and the exchange of information among the platforms and control centers is managed in data packages such as plots and tracks. These systems are designated as information technology (IT)-based network-centric operations [5]. The traditional applications of the platform-centric sensor concept are based on separately developed and maintained mechanical, electronic, and electrical engineering systems. As an example, the collocated information of the Secondary Surveillance Radar (SSR) and Primary Surveillance Radar (PSR) with separate or associated plot/track reports operate under a system of data fusion in the Command and Reporting Centers (CRC) only. The failure of the sensor or communication elements has been shown to reduce the data fusion efficiency, even in the case of redundancy and the use of modern data transfer networking technology. Recent military conflicts have demonstrated that reconfigurable and flexible applications, such as SDR (Software Defined Radios), satellite internet applications, embedded systems for measurement, and sensors & weapons on drone platforms related to I5.0 could be the most efficient solutions. The I5.0 model is equipped with state-of-the-art mass-produced sensor technology, which has the potential to be utilized in response to the rapid emergence of novel threats. Active sensing involves the generation of an excitation signal and the supply of such phenomena as RADAR, LiDAR, ultrasonic and SONAR in response to an external stimulus. This serves as the output signal, with no requirement for an additional energy source. Passive sensors, such as cameras, infrared, passive radar, and multi-spectral acoustic sensors, operate without the need for an external excitation supply and do not emit energy towards the target. Instead, they measure the natural energy that is reflected or emitted by the target. Smart sensing is defined by the execution of a predetermined function, frequently supported by AI applications, on a specific type of gathered data. This function is then transmitted through a networked connection. However, it is important to note that, again, the available manufactured IADS assets cannot be effectively used on the battlefield without military engineers who can adapt them to the rapidly changing requirements of the battlefield.

6.1.2 TOPIC-RELATED HUNGARIAN HISTORICAL ROOTS FOR EDUCATION

The focal point of environmental sensing is determined by the operational wavelength of active/passive sensors. The Moon Radar measurements, managed by Zoltán Bay, were carried out on February 6, 1946, two weeks later than the USA Diane project. The Hungarian Moon radar in the very high frequency (VHF) band worked at 121 MHz. It was fabricated on SAS military air surveillance radar antenna elements and subsystems [6]. Microwave holography, invented by Dénes Gábor in 1947 (1948?) and awarded Nobel Prize, is commercial off-the-shelf (COTS) technology today [7]. John von Neumann was a major contributor to mathematics, game theory, physics, economics, computing, and statistics [8]. Ferenc Krausz, who got the Nobel prize in 2023, generated attosecond pulses that allow amplitude and phase synchronization of sensor systems achievable to values close to 10^{-18} s [9].

Mechatronics, also referred to as mechatronics engineering, can be defined as a symbiotic approach to mechanical engineering and an interdisciplinary branch of engineering, including military, that has expanded upon the integration of mechanical, electronics, power electronic, computer science, robotics, telecommunications, systems engineering, AI control, and product engineering and electrical engineering systems. The extended synergy related to opportunities and challenges like the shortcomings of teachers and professors; refer to Fig. 6.1 introduced by Gy:orgy Ábrahám at the Hungarian Academy of Sciences Mechatronics Subcommittee in 2024 and under publication [10].

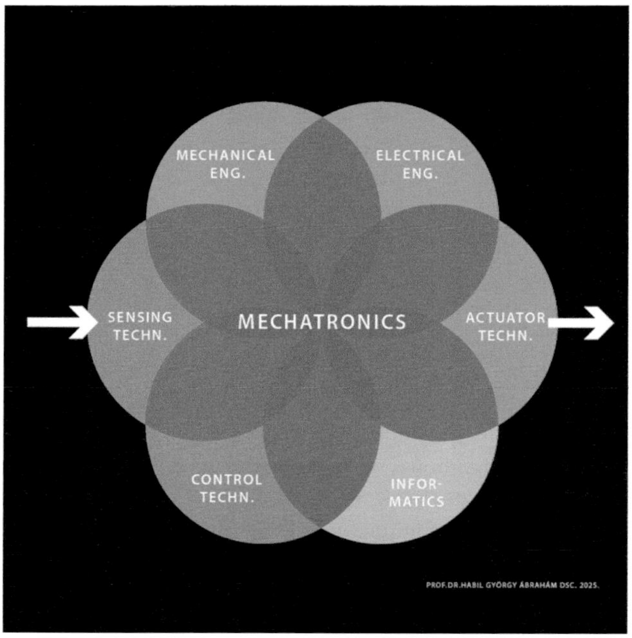

Figure 6.1 Star of mechatronics component (the "lotus flower").

The fundamental principle of mechatronics is to utilize a modeling approach to address the diversity of a product across multiple disciplines. Referring to Fig. 6.1, the following observations can be made. The modeling effort is required at different levels, including conceptual design, behavior-based design, technology selection and implementation, and manufacturing. From a mechatronics perspective, radar instruments are distinguished by their remarkable flexibility and multi-application life cycle capabilities. The modeling of main components, such as transmitters, power distribution units, chassis and steel structures, motion control units, indoor climate control (heating and cooling) units, radio frequency (RF) units and antennas, signal processing computers, and radar environment can be facilitated by a mechatronics approach. The multi-domain mechatronics modeling approach could be extended to utilize adaptive and cooperating machines for real-time parameter estimation during field operation.

This study examines the newly emerged threat sensing challenges for cost-efficient mechatronics engineering design solutions perspectives because the integration and interoperability of the sensors are vital elements to connect each layer of this architecture.

6.2 THE KEY NON-CONTROLLED CHALLENGES OF RADAR DESIGNERS

As illustrated in Fig. 6.2, the square-marked parameters represent those factors that are not within the radar designer's control and are crucial in determining the received energy and the detectability factor of the target. The key parameters, whose selection is at the designer's discretion, are marked with circles; these are: the operational wavelengths determine all key sensor performance behavior related to the target type and operational environment. There are well-developed solutions for transmit-receive chains even in the case of heavy jamming conditions. At the same time, the correlation coefficient-related challenges have started to be opened for public hot topic applications in the last decade only [11]. The realization challenges of high correlation coefficients, such as time delay differences among main and other channels as a function of signal bandwidth, the sensitivity of channels, gain differences among antennas, main antenna to other antenna sidelobe pattern mismatch, and speed of adaptively as a function of cross-correlation time settings are reducing the achievable values [12, 13]. The required complex cross-correlational coefficient gain should be rich in the case of 20 dB or 30 dB signal integration gain to 0.995 and 0.9995, respectively. However, "Why do we have two eyes and ears?" is permanent and helps us understand advanced radar design challenges.

The "Jammed Receive System Noise Temperature" module is the strongest factor that is not in the hands of the radar designer because devices such as Digital Radio Frequency Memory (DRFM) on the drone platform extend the freedom of jamming techniques [14, 15]. The "Receive Gain B" module is the sum of the received and integrated signals of the system. The R^2_{Rec1} and R^2_{Rec2} markings indicate the Quasi-Monostatic (Q-M) ranges among the receivers. These parameters should be evaluated by traditional radar design concepts to facilitate a more profound comprehension,

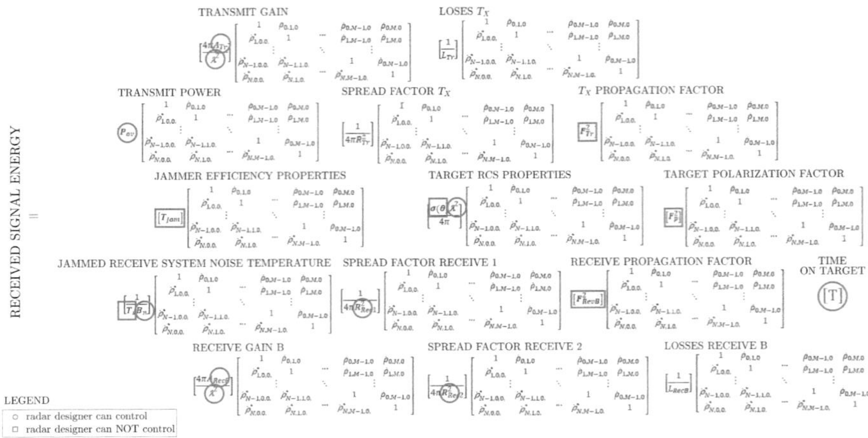

Figure 6.2 Radar performance concerning the environment, such as radar cross-section and jamming.

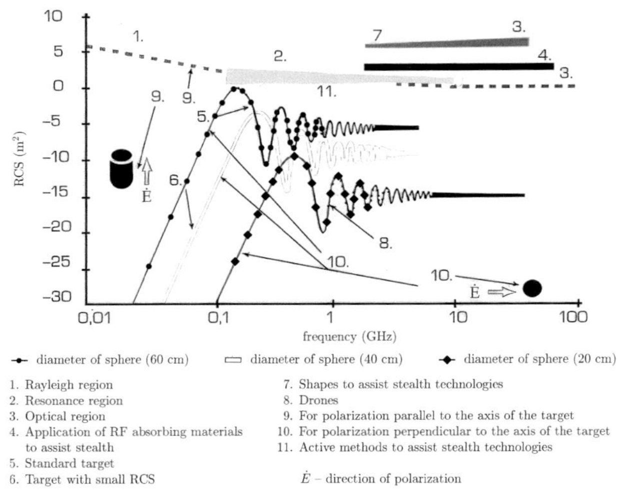

Figure 6.3 Radar cross-section characteristics with its countermeasure efficiency.

while it is acknowledged that there are concepts such as "tracking before detection" or "recognition before detection or tracking".

6.2.1 CHALLENGES OF TARGET DETECTION

The low RCS is pivotal to the "to be seen or not to be seen" performance. The RCS has been comprehensively evaluated in the extant literature [16], including by the authors themselves [17]. Consequently, Fig. 6.3 provides a concise summary of its

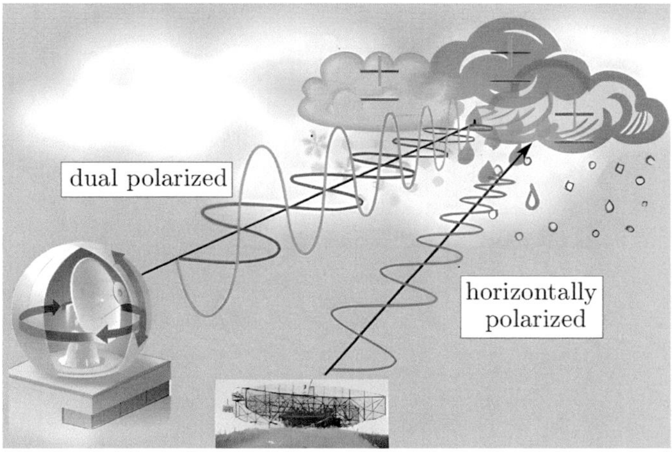

Figure 6.4 Illustration on one and dual-polarized radar operation (modified by authors [17]).

salient characteristics without the necessity for detailed elucidation. The majority of contemporary radar systems operate using a single polarization. However, I5.0-related MIMO (Multiple Input Multiple Output) or Q-M software-based and AI-supported solutions allow for the exploration of two polarizations simultaneously while also ensuring cost efficiency. It is important to note that the polarization factor offers advantages in radar design, a factor which was frequently not exploited in the past.

The propagation factors for transmission and reception are also strongly dependent on the radar operation wavelength and are now precisely calculated for different weather conditions. This is in contrast to the challenges faced by radar designers in the past.

As illustrated in Fig. 6.4, dual-polarized radar, like Multifunction Phased Array Radar (MPAR) Advanced Technology Demonstrator (ATD)-USA, offers

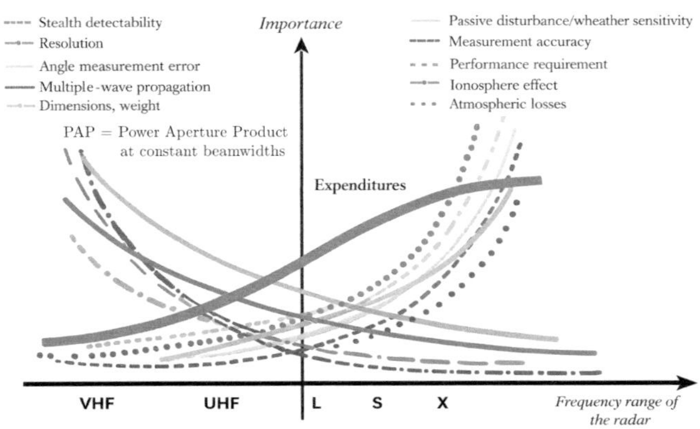

Figure 6.5 Different radar parameters by frequency range (modified by authors [17]).

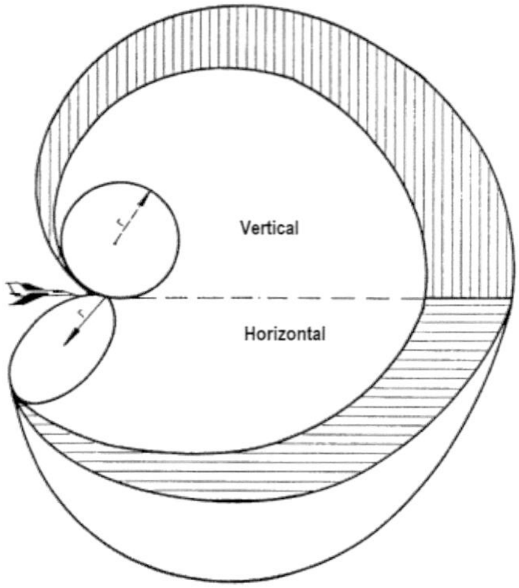

Figure 6.6 Maneuvering the performance of an aeroplane in vertical and horizontal directions [18].

distinct advantages, while Fig. 6.5 demonstrates the frequency dependence of radar characteristics' importance.

The electronic warfare performance of the radar has been identified as a significant game-changer on modern military battlefields (see Fig. 6.2). This is evidenced by the increasing radar receive system noise temperature and the concomitant reduction in active radiation thought to transmit side lobes. The raw order of magnitude measurement or simulation of several radar characteristics, including the propagation medium's RCS, polarization, loss factors, and signal strength reductions among signal paths, can be determined using Blake chart calculations.

6.2.2 CHALLENGES OF TARGET TRACK INITIALIZATION AND TRACK MAINTENANCE

When the target is detected with the required qualities, the detection probability with false alarm rate probability of the target type, the plot extraction, and track initialization processes are initiated.

The efficiency of the most common tracking algorithms, such as Kalman, multihypothesis, or particle filters, is determined by the quality of their input data. It could be improved by knowing the kinematic properties of the targets in advance, such as the position of gravitational mass, speed, and acceleration, which are strongly determined by the mass and maneuvering performance of the target (see Figs. 6.6 and 6.7) [18].

A comprehensive investigation into the kinematic properties and their dependence on the correlation coefficient can be found in [19, 20]. Fig. 6.8, developed by the

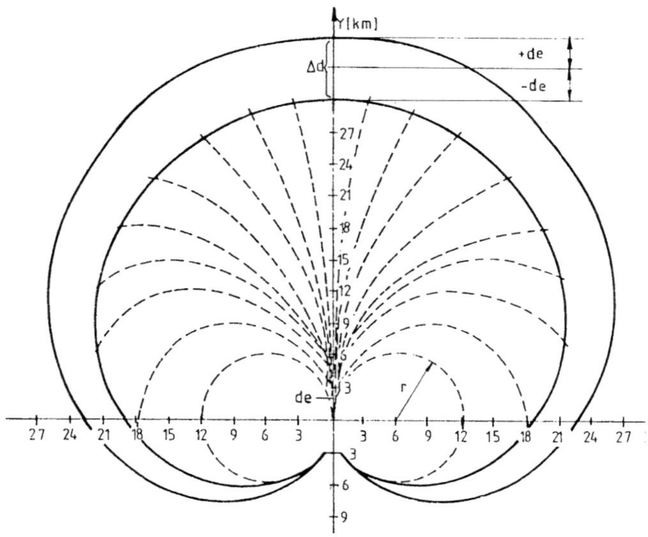

Figure 6.7 Maneuvering characteristics of flying objects from their speed [18].

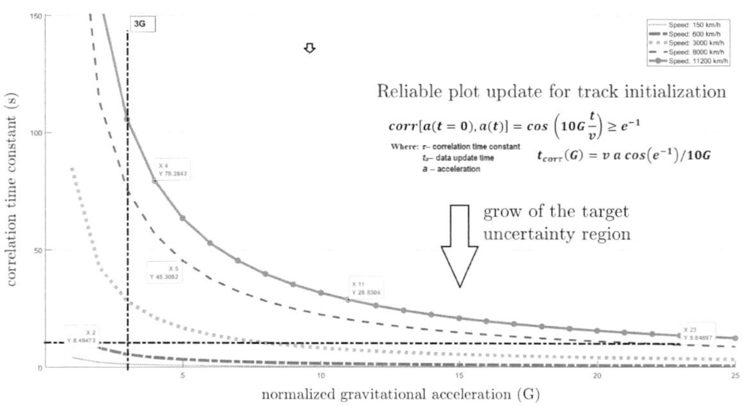

Figure 6.8 Track initialization requirements of the maneuvering targets (HyW by a solid line with circular markers).

authors, summarizes this relation with extension to hypersonic weapons (HyW). It is shown that traditional tracking algorithms could solve the HyW tracks' initialization and track maintenance after their reliable detection, while the track initialization and track maintenance of drones are more challenging.

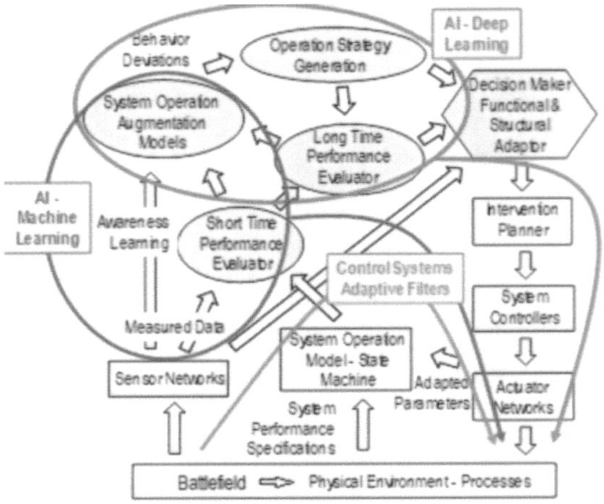

Figure 6.9 Advanced AI-supported sensor/radar of battlefield (modified on [24]).

6.2.3 CHALLENGES OF I5.0-BASED AI-SUPPORTED IADS APPLICATIONS

As illustrated in Fig. 6.9, the advanced AI-supported sensor/radar network (Internet of Things—IoT) structure of the battlefield is intricate, which determines the present and emerging challenges in the field of IADS relations [21, 22]. All activities and processes are triggered by the physical environment, such as the surveillance volume of the radar system, and the expected IADS effect is also realized here in the physical environment.

The data and information provided by the sensors are transmitted simultaneously in three directions. The first is the immediate decision-making module, the second is the module for short-term performance evaluators, and the system operation augmentation module (SOAM) processes recent events. Information from the short-term data processing module is transferred to the long-term information processing module with feedback to SOAM [23]. The operational settings of the various modules are determined by the system performance specifications managed by skilled operators [24]. The SOAM analyses the received data, extracts the pieces of information required for system performance optimization, and transfers it to the module responsible for the operation strategy efficiency. Based on incoming information, the decision-maker block determines the required real-time intervention, which shall be carried out by system controllers and actuator networks such as air defense guns, jammers such as DRFMs, interceptors, and missiles.

6.3 DRONE DEFENSE CHARACTERIZATION

The increasing prevalence of drones poses significant challenges to the protection of IADS assets and even critical infrastructure in peacetime. As illustrated in Fig. 6.10, a traditional stacked beam long-range radar (LRR) and its upgraded version, which

Figure 6.10 Vertical coverage of traditional radar short-range (yellow) and long-range detection (blue) [17].

is suitable for increased probability tracking of high-priority targets, are depicted. It shows the cosecant squared pattern (CSP) of the air surveillance volume of the LRR and also shows the time on target (ToT) challenge for the short-range radar (SRR) pattern vs. the LRR coverage. The LRR beams of the currently operational radars can detect targets up to a range of 460 km (250 nautical miles) and an altitude of 33 km (100,000 feet), with detection commencing from a range of 10 km. Consequently, newly implemented measures against drones between 0.1 and 20 km are required.

The integration of algorithms of interferometry, microwave holography, and polarimetric sensors/radar within advanced 3D sensors has been demonstrated to enable approximately 10 times higher precision in remote sensing, real-time environmental monitoring, and enhanced situational awareness in comparison with traditional systems [25]. To illustrate this point, the key factors of the 24 GHz mass-produced automobile and the unique Hungarian Moon radar main properties and performances are prepared by the MATLAB radar designer toolbox for demonstration, shown in Table 6.1. After target extraction, the remaining radar system design tasks must be addressed. The subsequent challenges pertain to target initialization and track maintenance. Once a reliable target plot has been extracted and the related track processed by the system, any additional customer requirements can be addressed, including target recognition, engagement, environmental clutter, and interference mapping. It is important to note that all of these processes are supported by AI.

It can be concluded that the performance of the mass-produced automotive radars in Q-M operational mode fulfills the drone detection and tracking requirements within an approximately 3 km range. Despite their compact dimensions of approximately 50 × 50 mm, the radars provide the expected 300–400 m target detection range in clear weather conditions shown in Fig. 6.11. The primary limitation of the system is atmospheric scattering, which increases losses by several orders of magnitude in rain.

Table 6.1
Key Technical Characteristics of an Automobile and the Hungarian Moon Radar

Name	24 GHz	121 MHz	Comments
R_{max} (km)	0.385	76	$P_d = 0.95$; $P_{fa} = 10^{-6}$
Target RCS (Sw1) (dBsm)	0	0	
Bandwidth (MHz)	300	0.125	
Peak power	30 mW	20 kW	
Pulse width (μs)	50	8	
PRF (kHz)	20	0.75	256 pulse coherent integration
$G_{Tr} = G_R$ (dBi)	20.33	19.0	antenna efficiency $= \mu = 70\%$
Antenna surface (m^2)	0.0017	54.1	
Atmospheric model	standard	standard	weather condition: *clear*
Noise temperature (K)	8000	2750	

Note: Automobile radar for 76–81 GHz has similar performance, while the antenna surface is 3.25 times smaller.

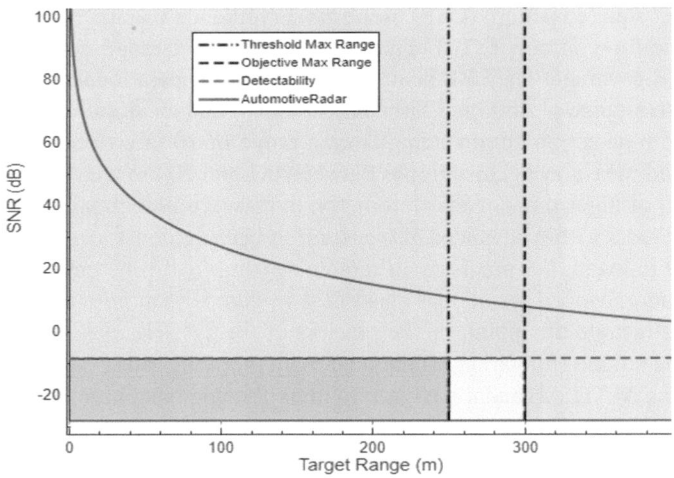

Figure 6.11 24 GHz automobile radar target detection performance in clear weather conditions.

6.4 HYPERSONIC WEAPONS CHARACTERIZATION

The HyW is defined as a class of long-range precision strike weapons that travel at a speed of at least Mach 5 or higher, operating at altitudes between 23 and 60 km. These weapons are distinguished by their low RCS. The material in [26] is of particular value, as it establishes the foundation of the basic objectives, financial burden, and effectiveness of HyW for consideration. Additionally, an overview of the historical progression of HyW projects is provided in [27].

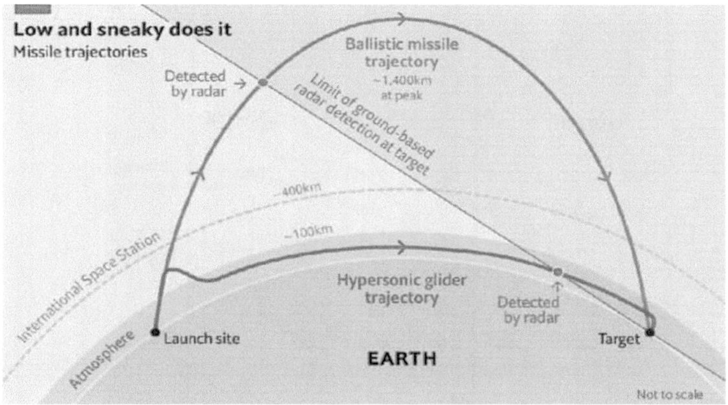

Figure 6.12 Key performances and the operational environment of the HyW [26].

6.4.1 THE KEY PERFORMANCES OF THE HYPERSONIC WEAPONS

There are two primary categories of HyW:

1. hypersonic glide vehicles (HGV) are launched from a rocket before gliding to a target;
2. hypersonic cruise missiles (HCM) are powered by a scramjet (note that supersonic combustion ramjet is a variant of a ramjet air-breathing jet engine in which combustion takes place in supersonic airflow).

In contradistinction to ballistic missiles [28, 29], HyW do not pursue a ballistic trajectory and are capable of maneuvering en route to their destination. The performance of HGV in the operational environment of the HyW is illustrated in Fig. 6.12, which compares the detection timelines of ballistic missiles and HGV by terrestrial-based radars. Consequently, space-based sensor systems appear to be a straightforward solution to hypersonic threats; however, they are single applications and require alternative solutions for redundancy in a cost-efficient manner.

As illustrated in Fig. 6.13, the velocity of HyW is compared with that of conventional aircraft to demonstrate the time limitations imposed by IADS. The HyW reference point is situated in the airspace of Budapest, Hungary. It is evident from the data presented in Fig. 6.13 that civilian aircraft can reach destinations such as Vienna or Belgrade. The modern Saab JAS-39 extends the range to Warsaw, while high-speed reconnaissance aircraft such as the SR-71 or MiG-25 can operate within the vicinity of Budapest and Kyiv. In the case of a 20-minute operational cycle, a contemporary HyW aircraft could travel the distance to Beijing, China. Fig. 6.14 illustrates the Cross-Section/Elevation (CSC) elevation coverage of a surveillance radar antenna pattern currently in operation and required against the HyW targets. An extension of the conventional radar detection coverage is imperative, encompassing a range of up to 1000 km and a minimum height of 60 km for HyW detection, tracking, and recognition capabilities.

Figure 6.13 Comparison of traditional aeroplane speed with the HyW speed developed by authors.

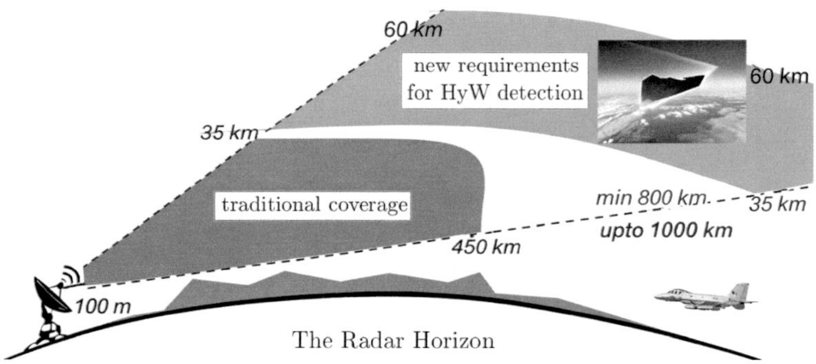

Figure 6.14 Vertical coverage of traditional radar (blue) and new requirements for HyW detection (orange).

A more in-depth investigation of the subject matter, primarily based on [26, 27] and Blake chart calculation [30], has identified the following key characteristics of HyW:

1. *RCS*—The capacity of HyW to repatriate energy to the receiving antenna is contingent on materials that demonstrate resilience to extreme environmental conditions, such as elevated temperatures and rapid velocities. The efficacy of the system is contingent on the wavelength, aspect angle, and signal fluctuation. The prevailing assumption is that for the S band (3.1 GHz), the RCS varies between 0.02 and 0.1 m^2; for the L band (1.3 GHz), it varies between 0.08 and 0.3 m^2; and for the VHF band (0.2 GHz), it varies between 0.2 and 1 m^2. The fluctuation model should be close to Sw1 or Sw3 as the generated plasma moderates the reflected echo signal fluctuation. As demonstrated in Figs. 6.3, 6.4, and 6.5, the shape and dimensions of HyW could contain a resonant region within the VHF band, exhibiting signal

polarization fluctuation as it maneuvers. However, the RCS determination is subject to uncertainty due to the variable plasma density surrounding the HyW, which is contingent on factors such as speed, maneuvers, and radar wavelength. All radar systems capable of radiating HyW from the upper side can circumvent this limitation.

2. *Operational altitude*—The launch should be conducted from ground, sea, and air platforms. It is assumed that the launch and the booster phase could be detected by vertically polarized long-range air surveillance radar and/or Low Orbit Satellites (LOS). The presence of Electronic Counter Measures (ECM) against ASS is also likely.
3. *Speed, maneuverability, and accuracy*—The object's velocity is measured around one kilometer per second, and it displays an exceptional degree of maneuverability, enabling it to operate effectively in environments where there is a high degree of contestation.
4. *Heat*—At hypersonic speeds, significant amounts of heat are generated due to the friction and air resistance experienced by the object in question. This heat can be detected by infrared sensors.
5. *Sound*—At hypersonic speeds, significant amounts of sound and low-frequency electromagnetic waves are generated due to the friction, plasma, and air resistance experienced by the object in question. This can be detected by acoustic and low-frequency sensor networks.
6. *Communication*—It is imperative to acknowledge that fundamental operations, such as communication, become a considerable challenge during hypersonic flight. To ensure the seamless connectivity required for effective operations, a system must be capable of maintaining communication links with both operators and decision-makers. This can be achieved through the utilization of LOS global communications and sensor systems. Consequently, the HyW could be considered as being vulnerable to advanced ECM from LOS.

6.4.2 THE ANALYSIS OF TRADITIONAL SENSOR/RADAR PERFORMANCES FOR HYW DETECTION

Blake Charts are predicated on radar equations [30]. To illustrate the potential, let us consider a scenario in which the maximum detection range of the L, S, and VHF band radar for "Standard Target" performance in a clear environment is 300 km. The radars operating in different frequency bands are designed in such a way as to have a total loss budget of 20.5 dB. The impact of the Frequency Diversities (FDs) mode of operation is also addressed. FD = 1, corresponding to a single frequency, and FD = 2, corresponding to two carrier frequencies. Table 6.2 provides a detailed characterization of the primary parameters of radar subsystems for the 300 km, RCS = 0.2 m^2 , Sw0, and $D_x(n = 4)$ = 12.63 dB, FD = 2 cases calculations. As demonstrated in the following cases, the performance of different radar systems and targets was investigated.

Table 6.2
The Main Performances of the Surveillance Radar Subsystems

Type	P_t	τ	$G_{Tr} = G_R$	F_x	T_s	$R_{\max}$
VHF	10^5	4.0×10^{-4}	23.3	2.55	31.15	300
L	4×10^4	2.0×10^{-4}	40	0.79	29.54	300
S	10^5	2.8×10^{-4}	40	0.79	30.01	300

Table 6.3
The Detection Performance of "Standard Target", Different Types of Aeroplanes and HyW.

$R_{\max}$ = 300 km,
in *clear* for RCS = 1 m^2, Sw1, P_d = 0.8 / P_{fa} = 10^{-6},
no Frequency Diversity, i.e., FD = 1

	L (1.3 GHz)		**S** (3.1 GHz)		**VHF** (0.18 GHz)	
Target type	FD = 1	FD = 2	FD = 1	FD = 2	FD = 1	FD = 2
RCS = 1 m^2, Sw 1	300 km	349 km	300 km	349 km	300 km	549 km
Airbus A320	531 km	619 km	447 km	522 km	791 km	923 km
Saab JAS-39	393 km	458 km	332 km	386 km	530 km	617 km
SR-71 / MiG-25	446 km	531 km	394 km	460 km	501 km	634 km
HyW	224 km	260 km	171 km	199 km	300 km	549 km

Note: Airbus A320 RCS = {VHF: 50 m^2; L: 10 m^2; S: 5 m^2}; Saab JAS-39 RCS = {VHF: 10 m^2; L: 3 m^2; S: 1.5 m^2}; SR-71 / MiG-25 RCS = {VHF: 8 m^2; L: 5 m^2; S: 3 m^2}; HyW RCS = {VHF: 1 m^2; L: 0.3 m^2; S: 0.1 m^2}.

As illustrated in Table 6.3, the calculation of the detection range of the L, S, and VHF band surveillance radar is demonstrated using the Blake chart for target types such as: the first target is defined as a "Standard Target" (RCS = 1 m^2, Sw1), with examples including the Airbus A320, the Saab JAS-39, the SR-71/Mig-25, and the HyW in a clear environment. The results obtained demonstrate the efficacy of lower frequency bands and the merits of frequency diversity when confronted with the challenges posed by HyW.

The general conclusion is that the traditional terrestrial-based radar network cannot detect hypersonic weapons until late in the weapon's flight.

6.4.3 THE ANALYSIS OF RADAR NETWORK TOPOLOGY SUITABLE AGAINST HYW

As a starting point, an update of currently in-production LRR systems is considered. Fig. 6.15 illustrates the vertical coverage of a multi-frequency band LRR, which has been demonstrated to be effective in the detection of ballistic missile trajectories and

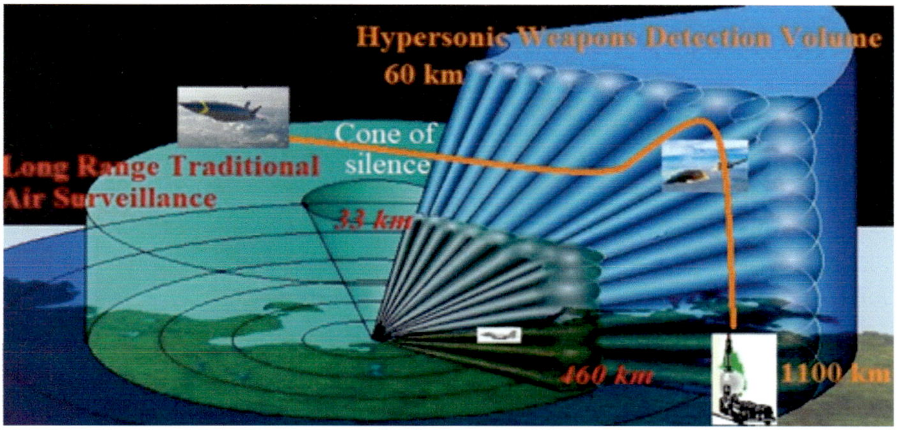

Figure 6.15 Traditional air surveillance radar and requirements against HyW.

may be used against various types of HyW. This solution incorporates L- or S-band LRRs, such as the AN/FPS117 and RAT-31 D/S, with the requisite new requirements for L, S, and VHF/UHF radar.

6.4.4 QUASI-MONOSTATIC SURVEILLANCE RADAR SYSTEMS BASED ON THE I5.0 PRODUCTION

Fig. 6.16 shows the operational principles of Q-M VHF, like P-18PL, radar topology for target surveillance. It is a state-of-the-art technology realization of spatially incoherent fusions of track-to-track fusion and plot fusion with video-signal fusion in the simple case. Short-term spatial coherence, such as plot fusion with radio-signal fusion, could be a cost-effective solution in heavy interference conditions. The spatially coherent signal fusion integrates matched amplitude and phase-correct signal sources of separated radars. It is acknowledged that the validity of the solution is contingent upon the multi-channel realization being the N-dimensional correlation matrix. The primary channel statistical relationship is to be maintained with all other channel correlation coefficients, as pointed out in Fig. 6.2. The correlation coefficients are delineated in Fig. 6.16 with the designations A11, A12, A21, and A22, consequently. It is important to note that this differs from the commonly applied correlation coefficients in that the phase of the two variables quantifies the strength of a linear relationship between band-limited RF signals. It is acknowledged that this presents a set of challenges for adaptive beam former/adaptive filter techniques.

The performance characteristics of the Q-M are determined by the doubled detection range (DDR) for low RCS as HyW, STAP (Space-Time Adaptive Processing) with super-resolution determined by baselines among the radars, and extended EPM (Electronic Protection Measures) with dedicated operator interactions. The rationale behind the DDR of the Q-M radar configuration is predicated on the augmentation of the transmit power by 3 dB, the transmit and receive antenna gain by 3+3 dB,

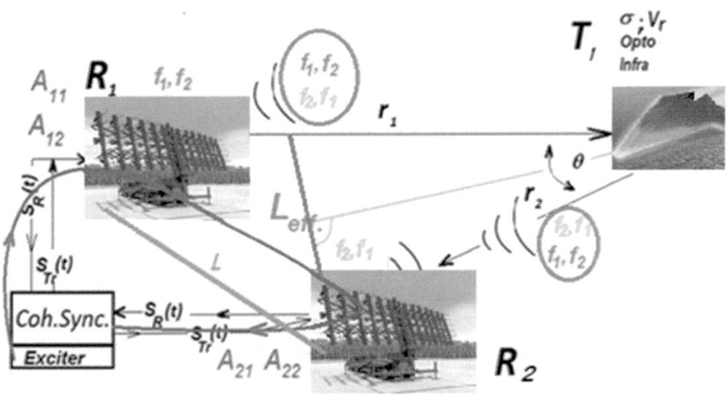

Figure 6.16 Operational characteristics of the Q-M VHF band radar system against HyW.

and double hits/echoes from targets, which collectively augment the traditional signal processing gain by 3 dB. The total increase in gain is equivalent to 12 dB. The VHF band necessitates particular consideration since it offers the RCS enhancement on 7–10 dB compared to the same types of S-band targets. Furthermore, a 2.3—2.7 dB multipath effect can be extracted from reflection on a flat ground surface [31]. Fig. 6.17 demonstrates the target detection performance of RCS = 0.2 m^2 for reduced detection probabilities, and extracted multipath effects are considered. A detailed analysis of the advantages of the Q-M radar configuration and a description of the parameters can be found in [32].

The challenges associated with achieving high values for the correlational coefficient are further compounded when attempting to establish a connection between two Q-M radars situated approximately 800 km from each other in Q-M operational mode on each side. This necessity arises due to the overlap of the vast cone of the silence of the VHF radar, which is more extensive than that of the S-band radar (see Fig. 6.15). To address this challenge, the implementation of Q-M radars in Poland and Hungary is proposed. These radars would serve to detect each other's cones of silence, as illustrated in Fig. 6.18.

6.4.5 Q-M-BASED VHF SURVEILLANCE RADAR NETWORKS FOR UNIVERSITY-LEVEL EDUCATION PROPOSE

A Q-M-based VHF radar for short-baseline and long-baseline applications are under installation at the University of Debrecen, Faculty of Engineering. Figure 6.19 shows the location and structure of the system. It was proposed in May 2024, and the RF units are using EMERSON-NI PXI workstations, test equipment of Rode & Schwarz, that imitate the Hungarian historical radar achievements in a modern industrial environment. The RF cabling to the roof from the laboratories is installed, but the required installation permissions for the antennas are pending.

The role of these sensors is high-level education training in which the student learns how the sensors are eliciting, collecting, and transforming data concerning

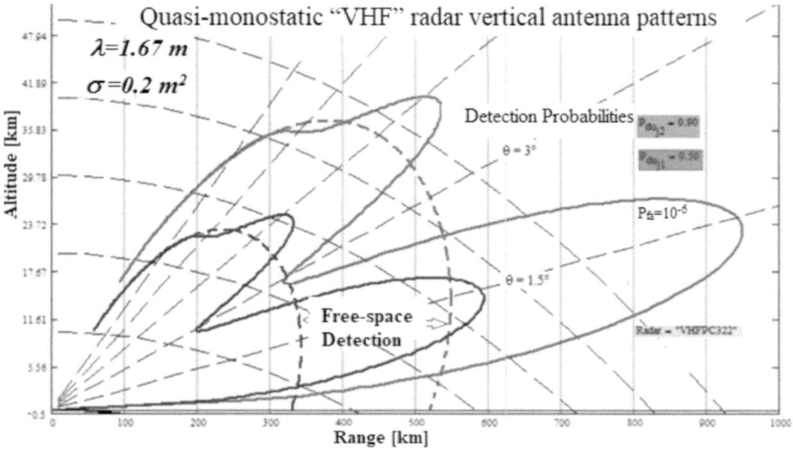

Figure 6.17 Vertical pattern of advanced Q-M VHF band radar.

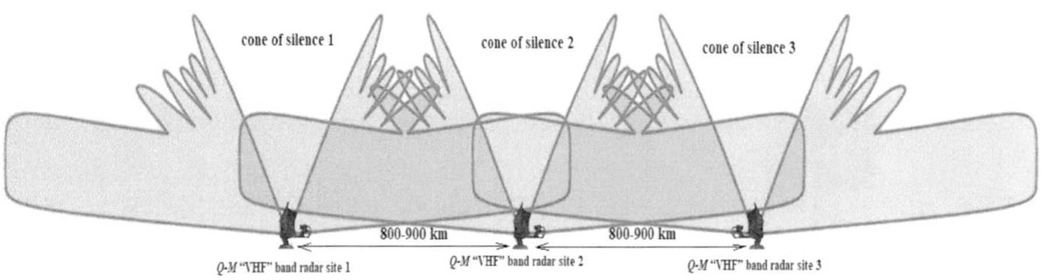

Figure 6.18 Vertical coverage of the Q-M VHF radar network with long baselines proposed against HyW.

the physical environment, including the system itself in case of heavy clutter and interference. Additional different types of sensors may be applied as individual detection means but also as elements of multi-sensor arrangements and sensor networks. The operational structure foreseen the collaboration with amateur radio “ham radio” clubs that have expertise for usage of the Earth-Moon-Earth (EME) or “moonbounce” path, which would extend students’ interest.

6.4.6 Q-M VHF SURVEILLANCE RADAR NETWORKS AUGMENTED BY PASSIVE RADAR

Q-M-configured VHF radar networks are enhanced with passive radar technology, which uses existing signals from other sources, such as commercial broadcast stations and/or communication signals, and/or other VHF radar signals, to detect and track HyW and drones [33, 34]. The augmentation provides several benefits, including stealth detection, covert operation, and cost-effectiveness due to the elimination of high-power transmitters [35, 36]. A topology of the solution is depicted in Fig. 6.20.

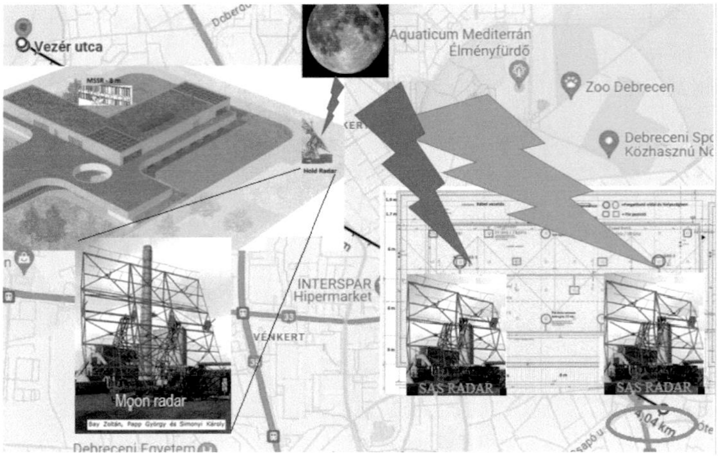

Figure 6.19 A topology of Q-M SAS radar, short baseline, and Q-M SAS to Moon radar, long baseline, 4.04 km, VHF radar network.

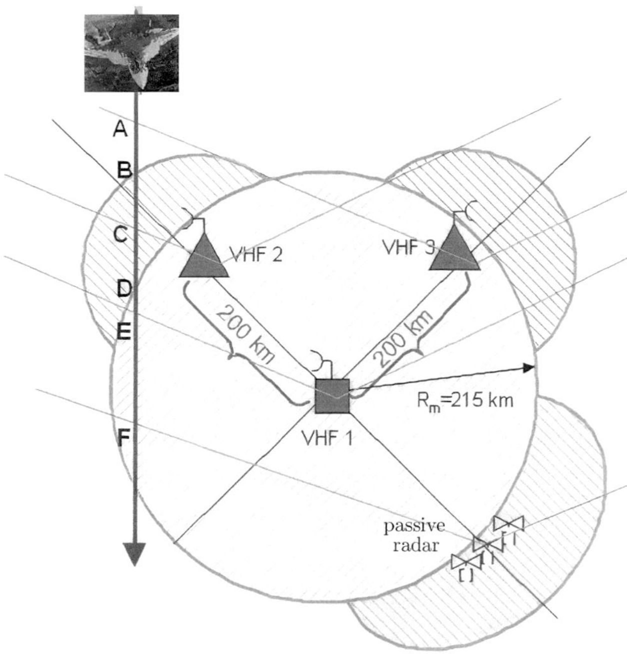

Figure 6.20 A topology of an operational VHF surveillance radar network augmented by passive radar.

6.4.7 MULTI-STATIC VHF SURVEILLANCE RADARS WITH EXPLORATION OF IONOSPHERE REFLECTION

The over-the-horizon backscatter (OTHB), also referred to as high frequency (HF) band radar, is a radar system that utilizes the unique characteristics of radio wave propagation on the sea surface, with a range of up to 150–200 km, and the reflections from the ionosphere, which can extend up to a few thousand kilometers. There are relatively new radar types that exploit reflections from the ionosphere and not OTHB [37]. The principal feature of this radar type is the status of the ionosphere, under circumstances where the intensity of solar radiation generates sufficient electron cloud density for HF, VHF, and sometimes for UHF radar echoes from the ionosphere. In case of an insufficient electron cloud above the operational region, a dedicated HF radar heats the ionosphere. There are already operational radar systems, such as the military Resonance-N [38] and the civilian EISCAT's 3D system [39].

The EISCAT 3D system is a civil multi-static radar system operating in the HF and VHF range. It was constructed through an international collaborative effort in the territory of Ramfjordmoen, near Tromso, Norway; Kiruna, Sweden; and Sodankylä, Finland. The system's primary technical parameters are thus accessible for analysis. Concurrently, military radar systems function on a similar principle, exemplified by the Russian Federation's Resonance-N radar system, which remains underresearched, yet can be comparatively analyzed by examining the physical characteristics of their operation. The significance of Figs. 6.21 and 6.22 lies in the fact that they illustrate the physical operation of the radar beams reflected from the most significant layer of the ionospheric radar. It is important to note that in multi-static mode, signals emanating from several radar sites a few hundred kilometers from

Figure 6.21 EISCAT 3D Kaiseniemi, Sweden (photo: Johan Svensson [39]).

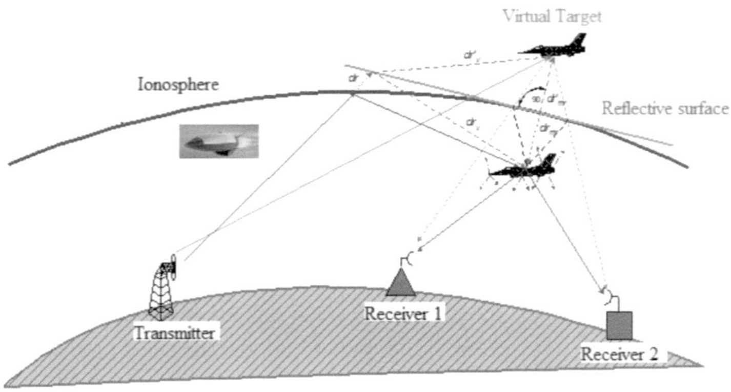

Figure 6.22 The topology of multi-static VHF surveillance radar systems with exploration of ionosphere reflection.

each other and reflected from the upper hemisphere of the stealth and HyW targets are received at each radar location, as Fig. 6.22 indicates [40, 41].

6.4.8 REALIZATION OF LOS Q-M CUBESAT-TYPE SURVEILLANCE SENSOR/RADAR SYSTEMS

The future growth of I5.0 sensor/radar technologies is contingent on several factors, including but not limited to: the miniaturization of sensors, the reduction of power consumption, the augmentation of measurement degrees of freedom, the simplification of calibration, and the enhancement of flexibility and conformability for usage [42]. The rapid pace of industrial development in the present age is inextricably linked to environmental pollution and degradation [43]. The pervasiveness of this issue has now reached a point where even the satellite environment in Earth's orbit is being contaminated. As a result, similar to airspace by air traffic above a given country, "space junk producer" companies and countries must pay fees for flying over a given LOS region. For this to happen, the first step is to verify that the position and trajectory of the space debris are measured. Concurrently, space-related engineering activities are undergoing exponential growth, as indicated in Fig. 6.23, coinciding with a notable increase in student interest. This phenomenon is of particular significance from the engineering education perspective, as it enables students to engage with subjects that align with their interests, thereby enhancing their learning experience.

Several papers have been published that propose solutions to the challenges posed by the increasing number of satellites in space, leading to a lack of space situational awareness (SSA) and cooperation in space debris mitigation [32]. CubeSats are satellites composed of standardized 100 mm boxes. The development of commercial-off-the-shelf (COTS) components has been a key advancement in recent years, leading to the promotion of CubeSats 1U due to their utilization of reduced size and weight,

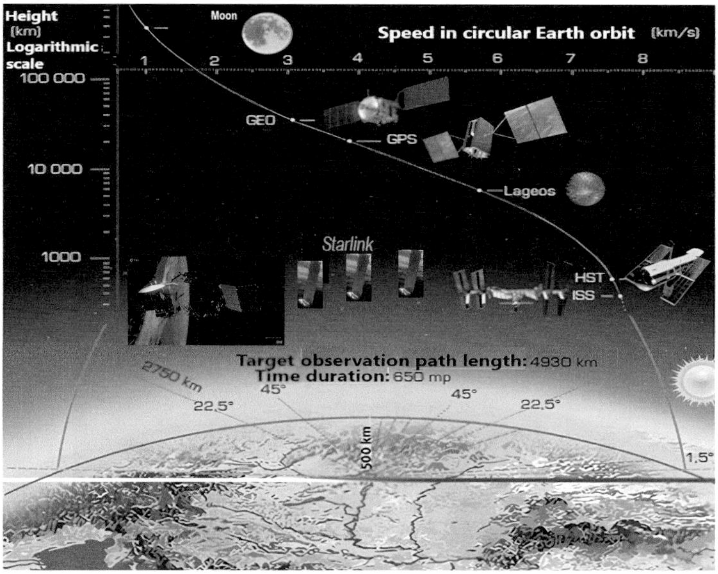

Figure 6.23 Space engineering types of equipment above the Carpathian Basin.

resulting in cost reductions and accelerated development times. RadCube is a "3-unit" CubeSat that has been recently developed for ESA by C3S in Hungary [44]. The "automobile" frequency band requires special attention because, in the case of the space offered, circumstances such as a "clear environment" without about 7000 K reduction of system noise temperature as demonstrated in Table 6.1. This factor increases the target detection range significantly, especially in passive radar operational mode, where one of the external transmit powers is related to Starlink satellites or the Sun itself. However, the reliance on COTS increases the susceptibility of nanosatellites to failures due to extreme environmental conditions encountered in space, such as ionizing radiation, temperature, and the Albedo effect. Analysis and modeling of the space Starlink Users' downlink for passive radar applications is available in [45].

Acknowledging the potential for resolving the technical challenges intrinsic to the development of CubeSat devices, we initiated an examination of the expectations and possibilities of their construction at the University of Public Service. Competition has been identified as an effective incentive to realize this concept, and students from interested universities are hereby challenged to maximize Q-M CubeSat performance shown in Fig. 6.24, within the following parameters:

1. Quantifiable and qualifiable measures serve to detect, track, initiate, maintain, and recognize space debris and/or meteor/small-size asteroids in low orbit and/or HyW above 40 km of the Earth's surface; objects are traveling with hypersonic speed. It is imperative to consider the exploitation of Starlink LOS in this context.

Figure 6.24 Q-M topology of space debris Smart CubeSat sensor in LOS (vision).

2. The initial sizes are 2 × (50 × 50 × 100 mm) and/or 2 × (100 × 100 × 200 mm). In the LOS orbit, the distance between two identical satellites is a minimum of 1 m and a maximum of 10 m.
3. It is imperative that both sets of components incorporate a minimum of two identical sensors operating with a coherent synchronizer/exciter within the following frequency ranges: 24 GHz, 76-81 GHz, infrared, and/or optical. It is imperative to note that the exploitation of Starlink LOS shall be taken into account for passive radar operation.
4. The devices must have locally built-in AI; the measured and pre-processed data must be sent down to the ground and receive control data from there. The exploitation of Starlink LOS shall be taken into account.
5. It is asserted that the system will demonstrate a reliable operational time in space for a minimum of three months.

6.4.9 NEAR-FUTURE IADS SENSORS VERTICAL REALIZATION VISION

The solution shown in Fig. 6.25 incorporates L- or S-band LRR, such as the AN/FPS-117, RAT31 D/S, and SMART-L, which is labeled with "L, S - Traditional" and "L - EXT." It also includes new requirements for L, S, and HF radar against HyW and missile trajectory detection. The region designated as "HyW VHF/UHF" signifies the recently mandated VHF/UHF radar surveillance coverage against HyW. The currently operational OTHB, HF band radar, is marked as "OTHB" and as yellow. Relatively large airspace, marked as "No Detection" (gray color), is indicated, and it is here that the implementation of new solutions such as ionosphere reflection exploration for HyW detection and tracking is required. The space debris sensor CubeSat

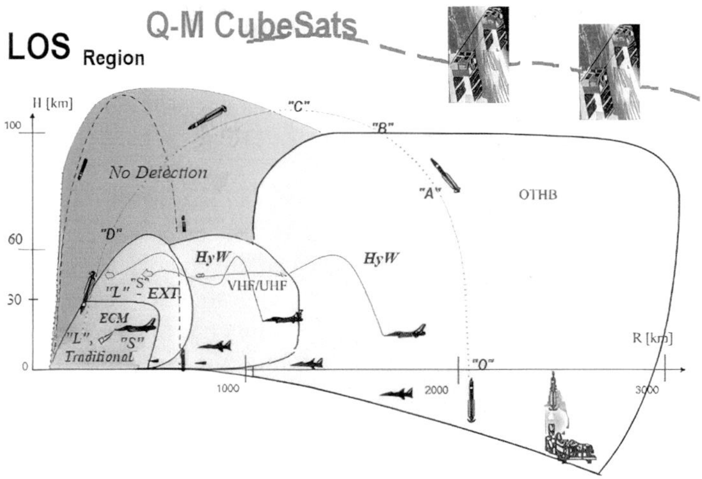

Figure 6.25 Operational range—high coverage of the multi-frequency band air surveillance radar network.

in LOS concerning "HyW detection and tracking" is introduced. The following are a few common realization aspects of this solution:

1. The basis for signal processing issues is based on two aspects due to the necessity for special T_x waveforms to facilitate coherent signal processing support in phase calibration, shift compensation, and adjustment of the propagation effect. Mission-critical waveforms are required for target detection, track initialization, track maintenance, and target recognition support. It is imperative to acknowledge that the technologies of electronic protection measures and electronic counter-countermeasures are founded upon advanced and cognitive signal processing algorithms. The implementation and maintenance of these technologies must be executed with the utmost precision and diligence.
2. *Radar Environment*—Propagation-related issues must be given due consideration, with particular attention paid to intensive impact from ground and ionospheric statuses. It is imperative to acknowledge the critical importance of the polarization and propagation properties of targets in this context. The mitigation of the radar horizon masking effect and the allocation of multi-frequency bands are prerequisites.
3. It has been demonstrated that the SRR and LRR performance fulfills the requirements for detecting small-sized RCS flying objects within a 20 km range. Real-world measurements have demonstrated the efficacy of the "Burnthought" operational mode of the radar in illuminating the COTS intruder drone.
4. *Electronic Warfare (EW) Condition*—EW activities are characterized by the joint use of the electromagnetic spectrum. In terms of planning, co-

ordination, and management, the spectrum is utilized from 1 MHz to 300 THz, incorporating the infrared and optical domains. The massive spread of DRFM devices is inevitable. The scope of electromagnetic spectrum operation has expanded to encompass network-connected, synchronized jamming devices, with the range of such devices extending from very short range to LOS.

5. It has been demonstrated that dual-polarized antenna beams, like MPAR ATD-USA, can address issues such as the focus of the beam, or beams, on dedicated surveillance sectors, as well as issues on the polarization of coherent signal processing gain and the very fast reallocation of the beam(s) positioning among prioritized targets. This is the functionality that is enabled by phased array radars with digital beamforming.
6. The processing of data pertaining to the initialization and maintenance of the HyW track necessitates the implementation of novel algorithms or the utilization of updated versions of preexisting algorithms [46].
7. The extension of radar network topology across Europe is a key development in the field of meteorological observation, while enhancing the surveillance performance of IADS. The radar coverage of a single radar site shall be extended by neighboring radar sites, which may be located hundreds of kilometers apart. The proposed structure will provide multiple overlapping coverage surveillance volumes and will address the gaps in radar coverage.

The author's opinion is that terrestrial-based radar networks with extended performances against HyW could fulfill these special requirements, but the task is still challenging.

6.5 CONCLUSIONS

This study underscores the imperative for enhanced performance, underpinned by I5.0/E5.0 principles and connected IADS against drones and Hypersonic Weapons. It emphasizes the necessity for the extension and exploration of the sensor/radar system measurements' degrees of freedom, which offers distinct advantages. The proposed solution integrates multi-sensor/radar and multi-static active-passive systems, comprising a Q-M topology connected to distributed command centers. The article goes on to depict the potential of a dual-polarized VHF radar being a focal part of near-future IADS. This "environmentally green" solution exploits the enhanced performance of merged, at the signal fusion level, and multi-band "coherent location systems" offered by AI supports. The central part of the nearby developed and implemented modern IADS is I5.0 mass-produced LOS Q-M CubeSats, 50 $\times$ 50 $\times$ 100 mm and/or 100 $\times$ 100 $\times$ 200 mm size devices for a multi-sensor fusion of active and passive systems. The initiative in Hungary challenges and competition on the "Moon radar-Hungary & Diana project-USA & Starlink" proposal shall be extended for NATO countries.

The article's objective is to stimulate additional development and discourse on the subject. The proposed temporary solutions are not yet fully developed and have the potential to be further refined.

AUTHOR CONTRIBUTIONS

Conceptualization, I.B. and L.S.; methodology, I.B. and L.S.; software, I.B. and L.S.; validation, I.B. and L.S.; formal analysis, I.B. and L.S.; investigation, I.B. and L.S.; resources, I.B. and L.S.; data curation, I.B. and L.S.; writing—original draft preparation, I.B. and L.S.; writing—review and editing, I.B. and L.S.; visualization, I.B. and L.S.; supervision, I.B.; project administration, I.B.; funding acquisition, I.B. All authors have read and agreed to the published version of the manuscript.

DATA AVAILABILITY STATEMENT

The authors confirm that the data supporting the findings of this study are available within the chapter.

FUNDING

This work received no specific grant from any funding agency in the public, commercial, or not-for-profit sectors.

REFERENCES

1. Mattes, P.W. 2019. What is a Modern Integrated Air Defense System. https://www.airforcemag.com/article/What-is-a-Modern-Integrated-Air-Defense-System/ (accessed March 12, 2025).
2. IDTechEx—Sensor Market 2025-2035: Technologies, Trends, Players, Forecasts. https://www.idtechex.com/en/research-report/sensor-market/1038 (accessed March 12, 2025).
3. IDTechEx—Quantum Sensors Market 2025-2045: Technology, Trends, Players, Forecasts. https://www.idtechex.com/en/research-report/quantum-sensors-market-2025-2045-technology-trends-players-forecasts/1063 (accessed March 12, 2025).
4. Galati, G., G. Pavan, and F. Daum. 2025. Lesson learnt from the rise and fall of quantum radar research. *Academia Quantum* 2(1). https://doi.org/10.20935/AcadQuant7586.
5. Weißß, M., S. Sandenbergh, F. Valdes, et al. Aspects of Next Generation Sensor/Radar Networks. In: Proc. of the 20th International Radar Symposium (IRS). Ulm (Germany), June 26–28, 2019, pp. 1–8. https://doi.org/10.23919/IRS.2019.8768188.
6. Balajti, I., and F. Hajdú. 2016. Surprising findings from the Hungarian radar developments in the era of the Second World War. *URSI Radio Science Bulletin* 358:82–108.
7. James, L. 2024. Casting a Light on Dennis Gabor: Inventor of Holography. https://www.allaboutcircuits.com/news/casting-a-light-dennis-gabor-inventor-of-holography/ (accessed March 12, 2025).
8. Dömölki, B. The double life of John von Neumann. In: Proc. of the 2023 IEEE 23rd International Symposium on Computational Intelligence and Informatics

(CINTI). Budapest (Hungary), November 20–22, 2023, pp. 153–158. https://doi.org/10.1109/CINTI59972.2023.10382125.
9. Maithani, S., A. Maity, W. Schweinberger, et al. Field-Resolved Infrared Spectroscopy using a Broadband Achromatic Interferometer. In: Proc. of the 2023 Conference on Lasers and Electro-Optics Europe & European Quantum Electronics Conference (CLEO/Europe-EQEC). Munich (Germany), June 26–30, 2023, pp. 1–1. https://doi.org/10.1109/CLEO/Europe-EQEC57999.2023.10232488.
10. Ábrahám, G., and I. Horváth. 2025. From multidisciplinary mechatronics to transdisciplinary intellectualized mechatronics. *IEEE/ASME Transactions on Mechatronics* (under final review).
11. Griffiths, H., and A. Farina. Multistatic and Networked Radar: Principles and Practice. In: Proc. of the 2021 IEEE Radar Conference (RadarConf21). Atlanta (GA, USA), May 07–14, 2021, pp. 1–5. https://doi.org/10.1109/RadarConf2147009.2021.9455149.
12. Chernyak, V. 1998. *Fundamentals of Multisite Radar Systems*. Amsterdam: Gordon & Breach Science Publisher.
13. Willis, J., and H.D. Griffiths. 2008. *Advances in Bistatic Radar*. London: Institution of Engineering and Technology.
14. Summerfield, J., and D. Kasilingam. Narrowband Bistatic Synthetic Aperture Radar Ambiguity Function Analysis and Design using Angular Harmonics. In: Proc. of the 2019 IEEE Radar Conference (RadarConf). Boston (MA, USA), April 22–26, 2019, pp. 1–6. https://doi.org/10.1109/RADAR.2019.8835524.
15. Baczyk, M., M. Wielgo, L. Maślikowski, et al. Multi-Static and Multi-Temporal ISAR Imaging of Non-Cooperative Air Targets. In: Proc. of the 2023 IEEE International Radar Conference (RADAR). Sydney (Australia), November 06–10, 2023, pp. 1–6. https://doi.org/10.1109/RADAR54928.2023.10371173.
16. Walenczykowska, M., W. Bużantowicz, and A. Kawalec. Impact of 3D Model Simplifications on the Determination of Numerical Accuracy of the Radar Cross-Section Values. In: Proc. of the 25th International Microwave and Radar Conference (MIKON). Wroclaw (Poland), July 01–04, 2024, pp. 254–259. https://doi.org/10.23919/MIKON60251.2024.10633938.
17. Balajti, I. 2020. *A Radar*. Budapest: Typotex (in Hungarian).
18. Kurucz, I. 1988. *A honi rádiótechnikai egység (magasabb-egység) harcvezetési és harctevékenységi folyamatainak korszerusítése, a vezetéstechnikai eszközök rendszerbe állítású-val összhangban*. Kandidátusi értekezés. Budapest: ZMKA (in Hungarian).
19. Ślesicka, A., A. Kawalec, B. Ślesicki, et al. Analysis of the Impact of Aircraft Maneuvers and Atmospheric Conditions on the Quality of Clutter Suppression in Space-Time Adaptive Processing. In: Proc. of the 2023 Signal Processing Symposium (SPSympo). Karpacz (Poland), September 26–28, 2023, pp. 152–157. https://doi.org/10.23919/SPSympo57300.2023.10302725.
20. Ösz, S. 1991. *A rádiólokátorok által felderített célok adatainak feldolgozása integrált automatikus rádiólokációs-információs rendszerrel (különös tekintettel a real-time adatfeldolgozás kritériumaira)*. Doktori értekezés. Budapest: MH Haditechnikai IntÃ©zet, 1-2 kÃ¶tet (in Hungarian).
21. Suto, J. 2024. Using Data Augmentation to Improve the Generalization Capability of an Object Detector on Remote-Sensed Insect Trap Images. *Sensors* 24:4502. https://doi.org/10.3390/s24144502.
22. Mark H.C., J. Botzheim, T. Thomessen, et al. 2020. Bacterial Memetic Algorithm Trained Fuzzy System-Based Model of Single Weld Bead Geometry. *IEEE Access* 8:164864–164881.

23. Suto, J. 2024. Improving the generalization capability of YOLOv5 on remote sensed insect trap images with data augmentation. *Multimedia Tools and Applications* 83:27921–27934.
24. Kocsis, I., B. BurjÃn-Mosoni, and I. Balajti. 2025. A Comprehensive Review of Key Cyber-Physical Systems, and Assessment of Their Education Challenges. *IEEE Access* 13:8898–8911.
25. Ślesicki, B., A. Ślesicka, and A. Kawalec. Analysis of the accuracy of the estimation of signal arrival angle in monostatic MIMO radar using the Capon algorithm and its modifications. In: Proc. of the 23rd International Radar Symposium (IRS). Gdansk (Poland), September 12–14, 2022, pp. 283–287. https://doi.org/10.23919/IRS54158.2022.9905010.
26. Sayler, K.M. 2025. Hypersonic Weapons: Background and Issues for Congress. https://crsreports.congress.gov/product/pdf/r/r45811 (accessed August 25, 2025).
27. Lockheed Martin Corporation. 2025. Revolutionizing Hypersonic Technologies. https://www.lockheedmartin.com/en-us/capabilities/hypersonics.html (accessed May 20, 2025).
28. Okoń-Fafara, M., and A. Kawalec. 2019. An analysis of the influence of the correlation gate shape on the TBD algorithm effectiveness. Proc. Volume 11442, Radioelectronic Systems Conference 2019, 114420D. https://doi.org/10.1117/12.2565751.
29. Walenczykowska, M., W. Bużantowicz, and A. Kawalec. Ballistic Missile Threat Modeling and VHF Radar Detection Performance Analysis for Tactical-Level Air Defense Simulator. In: Proc. of the 2024 IEEE International Workshop on Technologies for Defense and Security (TechDefense). Naples (Italy), November 11–13, 2024, pp. 385–390. https://doi.org/10.1109/TechDefense63521.2024.10863732.
30. Barton, D.K. 2007. *Modern Radar System Analysis Software and User's Manual*. Boston: Artech House, Inc.
31. Wu, J. 2020. *Advanced Metric Wave Radar*. New York: Springer.
32. Krebs, R., M. Andres, and M. Bockmair. A Modular and Scalable Architecture for RF Payloads on MALE RPAS. In: Proc. of the 20th International Radar Symposium (IRS). Ulm (Germany), June 26–28, 2019, pp. 1–9. https://doi.org/10.23919/IRS.2019.8768166.
33. Kulpa, K., M. Malanowski, M. Baczyk, et al. Hypersonic and Space Target Echo Modeling for Multistatic Passive Radar. In: Proc. of the 22nd International Radar Symposium (IRS). Berlin (Germany), June 21–22, 2021, pp. 1–8. https://doi.org/10.23919/IRS51887.2021.9466206.
34. Fränken, D., T. Ott, S. Lutz, et al. 2022. Integrating Multiband Active and Passive Radar for Enhanced Situational Awareness. *IEEE Aerospace and Electronic Systems Magazine* 37(8):36–49.
35. Rutkowski, A., and A. Kawalec. 2019. A concept of a microwave seeker designed for an anti-radiation missile. Proc. Volume 11055, XII Conference on Reconnaissance and Electronic Warfare Systems 2019, 110550P. https://doi.org/10.1117/12.2524749.
36. Malanowski, M., K. Jedrzejewski, J. Misiurewicz, et al. Passive Radar Based on LOFAR Radio Telescope for Air and Space Target Detection. In: Proc. of the 2021 IEEE Radar Conference (RadarConf21). Atlanta (GA, USA), May 07–14, 2021, pp. 1–6. https://doi.org/10.1109/RadarConf2147009.2021.9455158.
37. Infrastruktur ved UiT Ramfjordmoen forskningsstasjon UiT. 2025. https://uit.no/infrastruktur/enhet?p_document_id=685673 (accessed April 20, 2025).

38. Humpert, M. 2019. Satellite Images Reveal New Russian Long-Range Radar in the Arctic, High North News. https://www.highnorthnews.com/en/satellite-images-reveal-new-russian-long-range-radar-arctic (accessed April 20, 2025).
39. EISCAT Tromso site—EISCAT Scientific Association. 2021. https://eiscat.se/about/sites/eiscat-tromso-site/ (accessed April 20, 2025).
40. Abratkiewicz, K., P. Samczyński, G. Mazurek, et al. Passive Radar Using a Non-Cooperative Over-the-Horizon Radar as an Illuminator—First Results. In: Proc. of the 2024 IEEE Radar Conference (RadarConf24). Denver, (CO, USA), May 06–10, 2024, pp. 1–6. https://doi.org/10.1109/RadarConf2458775.2024.10548923.
41. You, J., X. Wan, H. Ke, et al. Resonance-region Target Detection with Wideband VHF Radar. In: Proc. of the IET International Radar Conference. Xi'an (China), April 14–16, 2013, pp. 1–6. https://doi.org/10.1049/cp.2013.0460.
42. Liu, Y. 2025. NCSTP: A Benchmark Dataset for Non-Cooperative Space Target Perception. 2025. https://figshare.com/articles/dataset/ _b_NCSTP_b_b_A_Benchmark _Dataset_for_Non-Cooperative_Space_Target_Perception_b_/28606754 (accessed June 16, 2025).
43. Kowalski, E., D. Conway, A. Morris, et al. Multifunction Phased Array Radar Advanced Technology Demonstrator (MPAR ATD) Nearfield Testing and Fielding. In: Proc. of 2019 IEEE Radar Conference (RadarConf). Boston (MA, USA), April 22–26, 2019, pp. 1–4. https://doi.org/10.1109/RADAR.2019.8835837.
44. Hungary Today: Hungarian Satellite Successfully Completes Its Mission. 2024. https://hungarytoday.hu/hungarian-satellite-successfully-completes-its-mission/ (accessed April 20, 2025).
45. Gomez-del-Hoyo, P., P. Samczynski, and F. Michalak. Analysis of Starlink Users' downlink for passive radar applications: signal characteristics and ambiguity function performance. In: Proc. of the 2023 IEEE Radar Conference (RadarConf23). San Antonio (TX, USA), May 01–05, 2023, pp. 1–6. https://doi.org/10.1109/RadarConf2351548.2023.10149600.
46. Gomez-del-Hoyo, P., and P. Samczynski. 2024. Starlink-based passive radar for Earth's surface imaging: First experimental results. *IEEE Journal of Selected Topics in Applied Earth Observations and Remote Sensing* 17:13949–13965.

7 The Invisible Swarm
How to Combine the Radar and Visual Characteristics of UAVs to Minimize Their Recognition

Ksawery Krenc
Faculty of Mechatronics, Armament and Aerospace,
Military University of Technology, Warsaw, Poland

CONTENTS

7.1 Introduction 104
7.1.1 Challenges in Detecting Medium-Class UAVs 104
7.1.2 Combining Radar and Visual Characteristics 104
7.2 Materials and Methods 105
7.2.1 Research and Parameters 105
7.2.2 Experiment Methodology 106
7.3 Comparative Analysis 108
7.4 Simulation Scenarios 109
7.5 Exemplary Calculations 110
7.5.1 Assumptions 110
7.5.2 Time-Invariant Swarm Formation 111
7.5.3 Time-Variant Swarm Formation 112
7.6 Results of the Numerical Experiments 112
7.7 Discussion 113
7.8 Conclusions 114

DOI: 10.1201/9781003739234-9

7.1 INTRODUCTION

7.1.1 CHALLENGES IN DETECTING MEDIUM-CLASS UAVS

In contemporary military literature, considerable attention is devoted to the issue of neutralizing unmanned aerial vehicles (UAV) [1, 2]. For many years, classifications of methods for destroying UAVs have existed, depending on the means used [3–5]. In military strategy, there is a principle that countermeasures are developed in parallel with operational methods [6, 7]. Therefore, the topic of an "invisible swarm", referring to a UAV swarm unrecognized by available reconnaissance means from a research and technical perspective, is intriguing, and from a military standpoint, highly desirable.

Contemporary surveillance systems, whether military or civilian, employ various reconnaissance methods. Among the most popular are: radar, visual, thermal, acoustic, etc.

As evident, these methods leverage distinct physical phenomena to identify diverse characteristics of observed objects, such as quantity, shape, and kinematics. It is also noteworthy that the subject under consideration pertains to a drone swarm rather than individual units attempting to evade unnoticed observation, thereby introducing additional complexity. This complexity encompasses various engineering disciplines, including radar reconnaissance, visual observation, robust object kinematics state estimation, material knowledge, and more.

7.1.2 COMBINING RADAR AND VISUAL CHARACTERISTICS

Due to the multidisciplinary nature of the considered problem in order to construct an "invisible swarm", one should address its key characteristics, such as: agent shapes, covering materials, particular swarm formation, trajectories, and dynamic parameters.

The modern capabilities of detecting objects of very small size make constructing an "invisible object" an extraordinary challenge. Naturally, the probabilities of detecting and positively classifying UAVs by adversary means increase proportionally with the number of swarm agents. Therefore, in addressing the above issue, one cannot rely solely on selected object features, such as shape or materials, while ignoring others.

For this reason, it should be stated that in a holistic approach, all domains influencing the detection and classification of swarm agents should be considered. This includes modeling the process of swarm recognition by adversary means. Such modeling can serve as the basis for further application of model attack-type strategies.

However, this chapter focuses on a specific segment of the described works, discussing the fusion of two most significant (though not the only relevant) object characteristics: electromagnetic, realized through radar reconnaissance, and visual, realized through a video camera. According to the author, due to the diversity of features provided by these two types of reconnaissance, strategies for evading detection by a drone swarm may have contrasting characteristics, which formed the basis for the considerations presented in this chapter.

7.2 MATERIALS AND METHODS

7.2.1 RESEARCH AND PARAMETERS

In order to mathematically model the discoveries and assumptions described in the introduction, a recognition function can be defined as follows:

$$\text{Rec}(O_i) = f\left(\hat{X}_\iota(t),\, F(t),\, s,\, m\right) \tag{7.1}$$

where $\hat{X}_\iota(t)$ denotes state vector estimate, $F(t)$ denotes swarm formation, m denotes material of UAV, and s denotes shape of UAV.

The swarm formation variable $F(t)$ is of an attribute type, and its possible values come from a finite, predefined set of labels, for example:

$$F : t \rightarrow \{\text{“V”}, \text{“LINE\_ABREAST”}, \text{“DIAMOND”}, \text{“STACK”}, \text{“ECHELON”}\} \tag{7.2}$$

The values of the recognition function may be defined according to the formula below:

$$[L_i, C_i, P_i] = \text{Rec}(O_i) \tag{7.3}$$

where i denotes UAV index, i = 1, 2, ... n, n denotes number of UAVs within the swarm, L_i denotes localization of the i-th UAV, C_i denotes classification of the i-th UAV, and P_i denotes probability of correct classification of the i-th UAV.

Probability P_i can be expressed as the conditional probability of correctly classifying the i-th UAV, given the successful localization of that UAV, as follows:

$$P_i = P_i\left(C_i \vee L_i\right) \tag{7.4}$$

Considering that each type of observer conducts independent observations, it can be inferred that a fusion process will be necessary to integrate these two distinct recognition systems and to produce the resulting (total) recognition function:

$$\text{Rec}_T(O_i) = \text{Rec}_R(O_i) \oplus \text{Rec}_V(O_i) \tag{7.5}$$

After establishing the overall recognition function for both areas, the process of rendering the UAV swarm invisible can be conceptualized as an optimization scenario, with the recognition function serving as the objective function. Consequently, optimization can be defined as the minimization of the recognition function:

$$\min_{\hat{X}_\iota(t), F(t)} \text{Rec}_T\left(\hat{X}_\iota(t),\, F(t),\, s,\, m\right) \tag{7.6}$$

As the probability $P_i = P_i\left(C_i \vee L_i\right)$, by definition, encapsulates information about the classification and localization of the UAV, it is justified to confine the consideration of the recognition function to that specific component. Thus, the total recognition for a single observation source may be expressed as follows:

$$\text{Rec}_T \approx 1 - \prod_{i=1}^{n}\left(1 - P_i(C_i \vee L_i)\right) \tag{7.7}$$

It is important to note that the state vector variable $\hat{X}_\iota(t)$, introduced in Eq. (7.1), refers to the state vector estimated by the radars and visual observers. With the knowledge of the location of each agent, the swarm can estimate these values, provided that the locations of the observers are known to the swarm, which may be described with the following formalism:

$$\begin{cases} \mathrm{Rec}_R(O_i): & \left(L_R, \hat{X}_\iota(t)\right) \rightarrow \hat{X}_\iota^R(t) \\ \mathrm{Rec}_V(O_i): & \left(L_V, \hat{X}_\iota(t)\right) \rightarrow \hat{X}_\iota^V(t) \end{cases} \tag{7.8}$$

Based on the Eq. (7.7) and assuming that the fusion operation (7.5) is performed according to Bayesian Theory, the total joint recognition probability (for both types of recognition) may be expressed as:

$$\mathrm{Rec}_T \approx 1 - \prod_{i=1}^{n}(1-P_i) \cdot \prod_{j=1}^{m}(1-P_j) \tag{7.9}$$

where i denotes radar UAV index, $i = 1, 2, \ldots n$; j denotes video camera UAV index, $j = 1, 2, \ldots m$; n denotes number of UAVs, recognized by radar; and m denotes number of UAVs, recognized by video camera.

7.2.2 EXPERIMENT METHODOLOGY

From the variety of UAV formations, commonly utilized in military applications [8–10], the author has selected four that they believe are the most pertinent for the considerations presented herein. These are:

1. v-formation (illustrated in Fig. 7.1),
2. line (illustrated in Fig. 7.2),
3. line abreast (illustrated in Fig. 7.3),
4. stack (illustrated in Fig. 7.4).

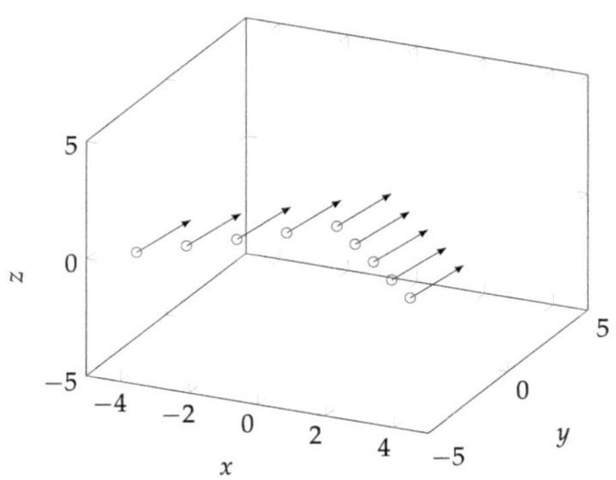

Figure 7.1 UAV v-formation.

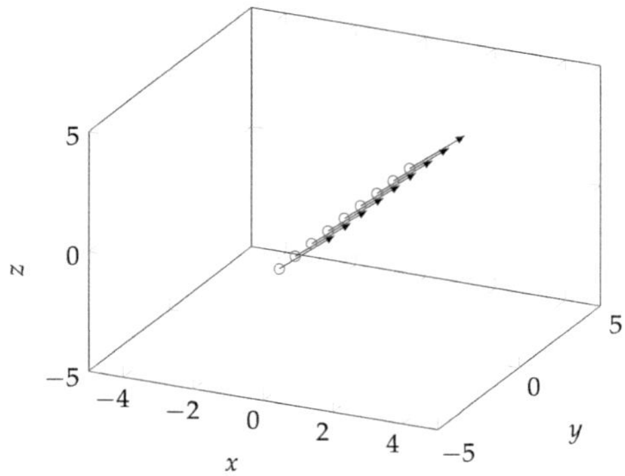

Figure 7.2 UAV line formation.

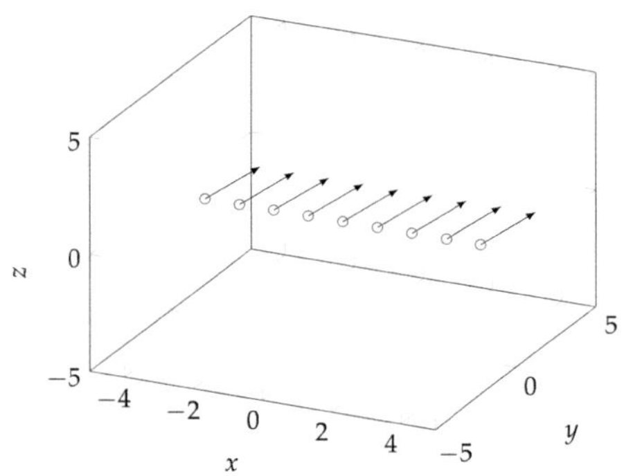

Figure 7.3 UAV line abreast formation.

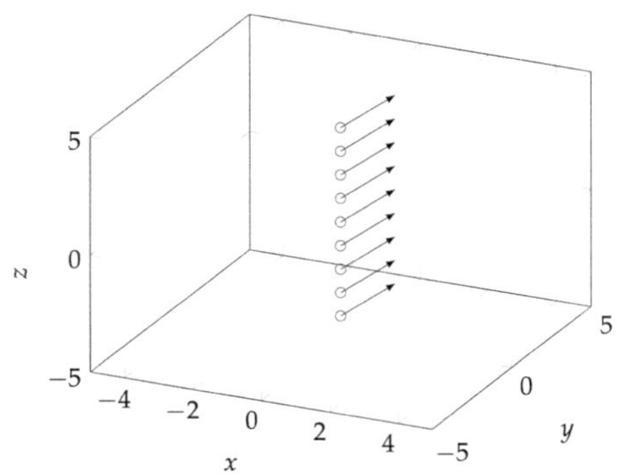

Figure 7.4 UAV stack formation.

These formations have undergone further analysis, calculations, and numerical experimentation, focusing on the tactical features of each formation and the dynamics of swarm agents.

However, the static components of the recognition function, such as the shapes of the agents or their surface material, have not been further considered. Thus, they serve as parameters in the recognition function, as defined by Eq. (7.1).

The selection of specific formations has been based upon the estimated radar and video recognition characteristics. For instance, the v-formation, inspired by natural v-flocks of birds, possesses distinct evasion potential, particularly concerning video recognition. Not only can it be challenging to distinguish from a genuine bird formation but certain visual aspects of the formation also obscure the true number of swarm agents.

The potential to obscure the number of agents can also be observed in formations such as the line and line abreast.

On the other hand, the stack formation boasts a significant radar evasion potential, which will be discussed in the following section.

7.3 COMPARATIVE ANALYSIS

Contemporary radar recognition enables the detection and tracking of various targets with high accuracy [11, 12]. Advanced radar systems employ advanced signal processing techniques and multi-mode operation to distinguish between different types of objects, including UAVs, aircraft, and ground vehicles [13–15]. However, one potential limitation of radar recognition lies in the resolution of elevation angles, which may result in challenges in accurately determining the altitude of targets, especially in cluttered environments or when targets are flying at low altitudes.

Regarding visual recognition, significant advancements have also been made in that area, driven by breakthroughs in artificial intelligence and deep learning [16–18]. Contemporary visual recognition systems utilize convolutional neural networks and other advanced algorithms to analyze images and identify objects with remarkable precision [19]. While visual recognition offers high-resolution imagery and accurate object identification, it may face challenges in certain scenarios, such as when swarm agents obscure each other, making it difficult to distinguish individual targets.

Considering the above, it is clear that different types of recognition have varying effects on the evasion potential associated with each type of recognition. This potential has been summarized in Table 7.1.

Upon initial examination of Table 7.1, it becomes apparent that, in terms of evasion, different formations are preferable depending on the specific type of recognition. There is no universally optimal formation for avoiding both radar and video detection. Thus, the exact swarm formation should be time-variant (which serves as a necessary condition) and, additionally, capable of rapid formation changes. This imposes further requirements on the swarm regarding the maneuverability of its agents.

Table 7.1
Potential of UAV Formations for Achieving Evasion of Radar and Visual Recognition

Recognition Type	**v-Formation** (V)	**Stack** (S)	**Line** (L)	**Line Abreast** (LA)
Radar	$n_R \to n$	$n_R \to 1$	$n_R \to n$	$n_R \to n$
Video	$n_R \to n/2$	$n_R \to n$	$n_R \to 1$	$n_R \to 1$

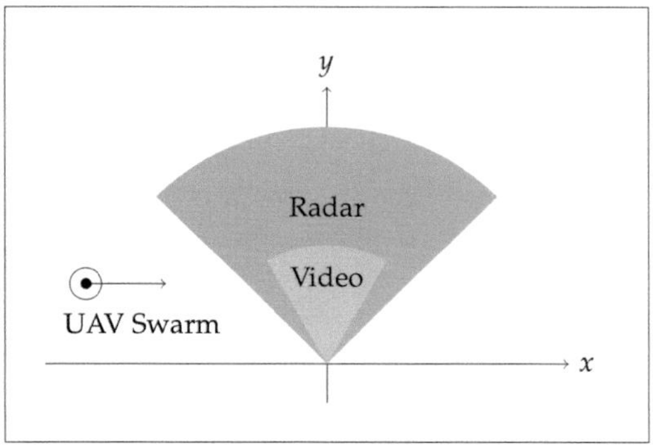

Figure 7.5 Scenario 1: Crossing the overlapping sectors of radar and video recognition means.

7.4 SIMULATION SCENARIOS

Taking into account the findings mentioned in the previous section, a series of five scenarios has been defined to scrutinize the possible movement of the swarm agents and their potential recognition, provided that the swarm has knowledge about the type and location of these recognition sources.

These five scenarios take into account both the configuration of the recognition sources and the approach of the swarm to the observers.

Regarding the configuration, two cases have been considered: overlapping observers' coverage regions and disjoint observers' coverage regions. Figures 7.5 and 7.6 present a situation where the sensors' recognition regions are overlapping, while Figs. 7.7–7.9 present the situation where these regions are disjoint.

Regarding the movement of the swarm, two cases have also been taken into account: crossing the source coverage region and heading towards the source.

It is not difficult to observe that in case of the disjoint region scenarios the recognition outcome may strongly depend on the fact whether the swarm is crossing the radar or video observer's region. Therefore both cases of crossing have been distinguished, which has been illustrated in Figs. 7.7 and 7.8.

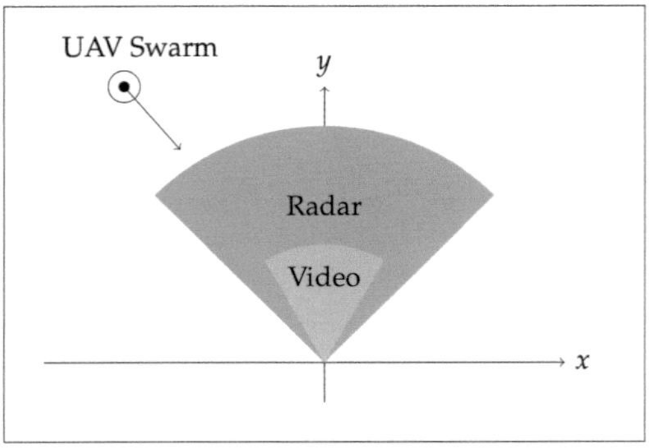

Figure 7.6 Scenario 2: Approaching the overlapping sectors of radar and video recognition means.

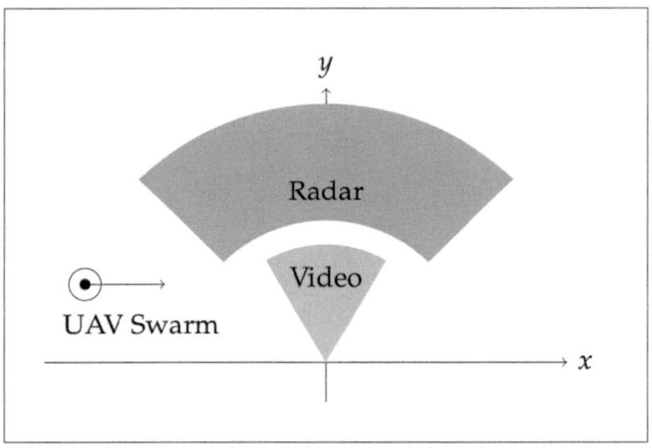

Figure 7.7 Scenario 3: Crossing the video recognition sector only.

7.5 EXEMPLARY CALCULATIONS

For the purpose of illustration of the considerations presented in this chapter, the author suggests to consider the following.

7.5.1 ASSUMPTIONS

Let us assume the swarm consist of the same type of the agents, and:

1. number of agents: $n = 9$;
2. location of recognition sources: L_R, L_V = *known*;
3. the total recognition probability for joint (radar and video recognition) is calculated according to Eq. (7.9);
4. classification probabilities of a single swarm agents for radar and visual recognition are equal: $P_R = P_V = 0.1$.

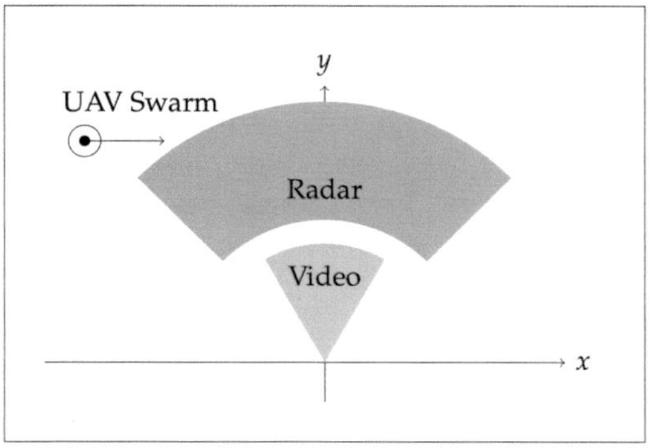

Figure 7.8 Scenario 4: Crossing the radar recognition sector only.

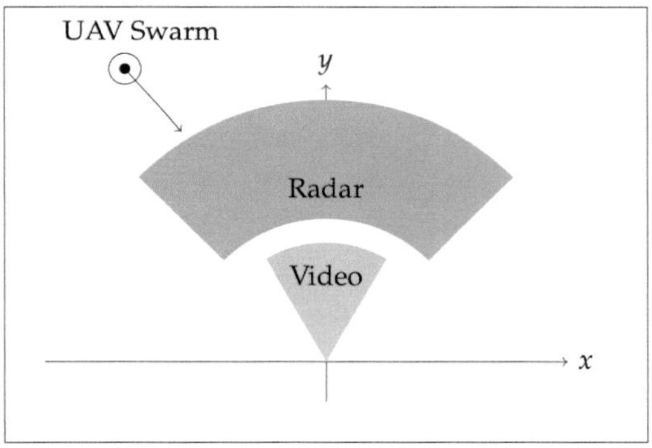

Figure 7.9 Scenario 5: Approaching the separate radar and video recognition sector.

Now, let us consider two cases referring to the swarm formation:

1. swarm formation is time-invariant: $F = const$;
2. swarm formation is time-variant: $F = F(t)$.

7.5.2 TIME-INVARIANT SWARM FORMATION

In the case of time-invariant swarm formation, each scenario has been analyzed with the assumption that the swarm formation remains unchanged over time. By considering the specific features of both types of recognition, as summarized in Table 7.1, one can calculate the resulting recognition probabilities, which are presented in Table 7.2.

For scenarios where the swarm passed through separate coverage areas, the resulting probabilities have been calculated for each area accordingly. The arrow indicates the order in which each particular area was passed.

Table 7.2
Recognition Probabilities Calculated for Each of the Scenarios—the Time-Invariant Case

Scenario	ν-Formation (V)	Stack (S)	Line (L)	Line Abreast (LA)
1	0.746	0.651	0.850	0.651
2	0.613 → 0.746	0.1 → 0.651	0.613 → 0.651	0.613 → 0.850
3	0.344	0.613	0.613	0.100
4	0.613	0.100	0.613	0.613
5	0.613 → 0.344	0.100 → 0.613	0.613 → 0.100	0.613

Table 7.3
Recognition Probabilities Calculated for Each of the Scenarios—the Time-Variant Case

Scenario	Sensor Area	Formation Sequence	Probability Sequence
1	$R \to R \cap V \to R$	S → LA → S	0.100 → 0.651 → 0.100
2	$R \to R \cap V$	S → L	0.100 → 0.651
3	V	LA	0.100
4	R	S	0.100
5	R → V	S → L	0.100 → 0.100

7.5.3 TIME-VARIANT SWARM FORMATION

In the case of time-variant swarm formation, each scenario has been analyzed with the assumption that the swarm formation could change over time to exploit the limitations of each type of recognition and pass through the observed areas unrecognized. The resulting probabilities have been presented in Table 7.3.

Given the assumed change in swarm formation, additional temporary probabilities can be calculated as the swarm passes through the observed areas. For instance, in Scenario 1, three zones (radar, radar + visual, and radar again) have been considered.

7.6 RESULTS OF THE NUMERICAL EXPERIMENTS

Analyzing the results of the calculations presented above, a couple of observations can be made.

The first, and most obvious, is that even with a very small single-agent classification probability of $P = 0.1$ in both cases (variant and invariant swarm formation), the resulting probability of recognition significantly increases with just nine agents. The fusion of these two sensors further emphasizes this effect. That makes the mission of the swarm to pass through the observed zones unrecognized very challenging to achieve.

The second important observation suggests that a time-variant swarm formation is capable of achieving much better performance compared to a time-invariant swarm formation. A changeable formation takes advantage of exploiting potential flows in

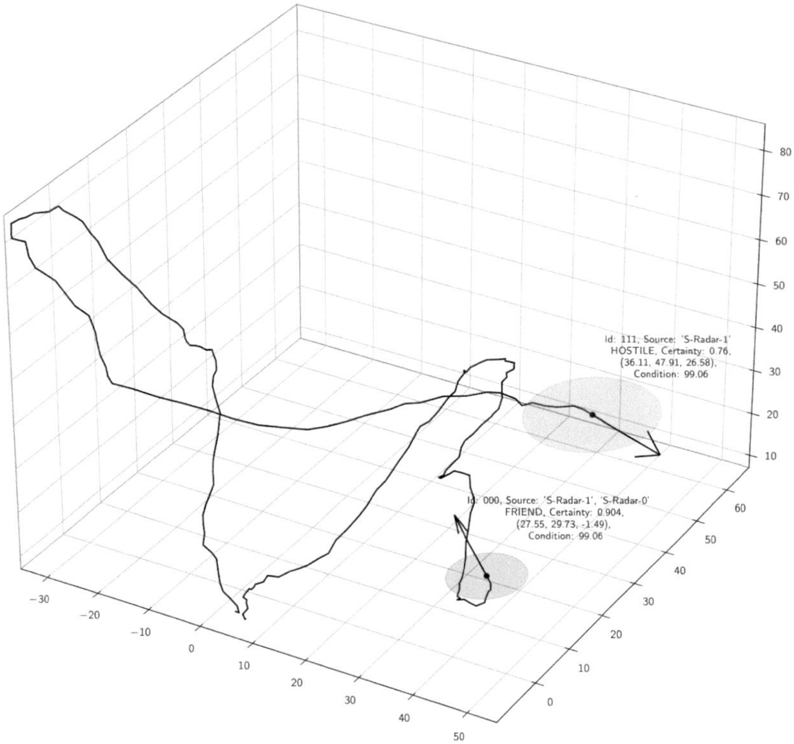

Figure 7.10 A view of the simulation scenario.

both types of recognition considered. Similar to how fusion amplifies the classification probability, the hybrid configuration emphasizes the evasion features of particular formations and, consequently, enables the reduction of those probabilities to a level that, in an adversary system, may be interpreted as clutter.

7.7 DISCUSSION

A critical reader might argue that if the considerations presented in this chapter were put into practice, the increase in classification probability resulting from the quantity of the swarm and the fusion of recognition means would be significantly greater than any potential reduction due to the changeable swarm formation.

In the author's opinion, the aforementioned points could hold true, especially considering that modern radar and visual systems are continually evolving. The vulnerabilities of recognition sensors, as demonstrated in this chapter, can currently be minimized. Additionally, it is important to note that exploiting these vulnerabilities requires a high level of precision, which may not always be achievable in all weather conditions.

However, the author also wishes to emphasize that the probabilities discussed pertain to the classification process occurring in the presence of clutter. In practical implementations, employing various techniques (which were not the focus of this study), such as camouflage techniques or reducing radar cross-section (RCS), may result in a lower detection probability. Nevertheless, the author believes that the ultimate goal is not to decrease the detection probability at any cost but rather to reduce the classification probability.

In order to illustrate that, a series of dedicated simulations has been conducted, incorporating specific sensor characteristics, precise UAV motion models, and information association and fusion algorithms. Figure 7.10 presents an example of such a simulation, where the certainty attribute indicates the classification probability assessed by a particular sensor.

7.8 CONCLUSIONS

In summary, the author would like to conclude that the key factor potentially constraining "invisibility", understood as the ability to remain unrecognized by radar and visual means, is not the mere detection itself but rather the positive classification of the swarm agents.

The classification of UAVs depends on multiple factors, some of which, such as the shape of the UAV or the coverage material, may be considered static. Their influence on invisibility typically occurs offline and is determined during manufacturing. Other factors, such as swarm formation or agents' kinematics, are considered dynamic. Thus, with knowledge about the locations of observation sources, these dynamic factors may determine the actual on-the-fly visibility of the swarm in specific weather conditions.

In the near future, the author plans to publish the results of the conducted exact model experiments, illustrating the concept presented herein.

DATA AVAILABILITY STATEMENT

The data that support the findings of this chapter are available from the corresponding author, K.K., upon reasonable request.

FUNDING

This work received no specific grant from any funding agency in the public, commercial, or not-for-profit sectors.

REFERENCES

1. Chamola, V., P. Kotesh, A. Agarwal, et al. 2021. A Comprehensive Review of Unmanned Aerial Vehicle Attacks and Neutralization Techniques. *Ad Hoc Networks* 111:102324.

2. Wojtyra, D., K. Wacławik, K. Krenc, et al. 2021. Concept for the Construction and Application of a Counter-UAV Defence System. *Problems of Mechatronics. Armament, Aviation, Safety Engineering* 12(1):87–100.
3. Bużantowicz, W., and P. Turek. 2020. Autonomous Combat Support Vehicles in Urban Operations: Tactical and Technical Determinants. *Advances in Military Technology* 15(1):97–114.
4. Rudys, S., A. Laučys, P. Ragulis, et al. 2022. Hostile UAV Detection and Neutralization Using a UAV System. *Drones* 6(9):250.
5. González, F., R. Caballero, F.J. Pérez-Grau, et al. Vision-based UAV Detection for Air-to-Air Neutralization. In: Proc. of the 2021 IEEE International Symposium on Safety, Security, and Rescue Robotics (SSRR). New York City (NY, USA), October 25–27, 2021, pp. 236–241. https://doi.org/10.1109/SSRR53300.2021.9597861.
6. Bużantowicz, W., and J. Pietrasieński. 2018. Dual-control Missile Guidance: A Simulation Study. *Journal of Theoretical and Applied Mechanics* 56(3):727–739.
7. Bużantowicz, W. 2024. An Airframe Stabilization System with State Observer and Correction Part for Terminal-Phase Missile Guidance. *Advances in Military Technology* 19(1):72–98.
8. Arquilla, J., and D.F. Ronfeldt. 2000. Swarming and the Future of Conflict. Santa Monica (CA, USA). https://www.rand.org/pubs/documented_briefings/DB311.html (accessed May 10, 2025).
9. Edwards, S.J.A. 2005. Swarming and the Future of Warfare. Santa Monica (CA, USA). https://www.rand.org/pubs/rgs_dissertations/RGSD189.html (accessed May 10, 2025).
10. Alkouz, B., and A. Bouguettaya. Formation-based Selection of Drone Swarm Services. In: Proc. of the 17th EAI International Conference on Mobile and Ubiquitous Systems: Computing, Networking and Services (MobiQuitous '20). Darmstadt (Germany), December 7–9, 2020, pp. 386–394. https://doi.org/10.1145/3448891.3448899.
11. Kawalec, A., and R. Owczarek. Specific Emitter Identification Using Intrapulse Data. In: Proc. of the 1st European Radar Conference (EURAD). Amsterdam (Netherlands), October 11–15, 2004, pp. 249–252. https://ieeexplore.ieee.org/document/1396531.
12. Filipiak, J., A. Kawalec, and E. Danicki. 1990. Wide-band SAW Dispersive Filter with a Flat Amplitude Response. *Ultrasonics* 28(6):355–357.
13. Kawalec, A. 2008. The Radiation Efficiency of SAW Periodic Strips Analysis at Higher Overtones Based on the Spectral Theory. *The European Physical Journal Special Topics* 154(1):117–121.
14. Ślesicka, A., and A. Kawalec. 2020. An Application of the Orthogonal Matching Pursuit Algorithm in Space-Time Adaptive Processing. *Sensors* 20(12):3468.
15. Walenczykowska, M., and A. Kawalec. 2022. Application of Continuous Wavelet Transform and Artificial Neural Network for Automatic Radar Signal Recognition. *Sensors* 22(19):7434.
16. Nguyen, P.H., K.W. Kim, Y.W. Lee, et al. 2017. Remote Marker-Based Tracking for UAV Landing Using Visible-Light Camera Sensor. *Sensors* 17(9):1987.
17. Du, L., C. Gao, Q. Feng, et al. 2017. Small UAV Detection in Videos from a Single Moving Camera. In: *Computer Vision. Communications in Computer and Information Science*, vol. 773, ed. J. Yang, et al. Singapore: Springer.

18. Liu, Y., L. Liao, L. He, et al. 2021. Trajectory and Image-based Detection and Identification of UAV. *The Visual Computer* 37(7):1769–1780.
19. Krenc, K., D. Gradolewski, D. Dziak, et al. 2022. Application of Radar Solutions for the Purpose of Bird Tracking Systems Based on Video Observation. *Sensors* 22(10):3660.

8 Numerical and Experimental Determination of the Radar Cross-Section Values for Air Defense System Performance Evaluation

A Comparative Study

Witold Bużantowicz

Marta Walenczykowska

Ksawery Krenc

Adam Kawalec
Faculty of Mechatronics, Armament and Aerospace,
Military University of Technology, Warsaw,
Poland

Branislav M. Ninković and Miodrag S. Tasić
WIPL-D d.o.o., Belgrade, Serbia

Branko M. Kolundžija
University of Belgrade, Belgrade, Serbia
WIPL-D d.o.o., Belgrade, Serbia

Rafał Przesmycki, Marek Bugaj, and Leszek Nowosielski
Faculty of Electronics, Military University of Technology,
Warsaw, Poland

DOI: 10.1201/9781003739234-10

CONTENTS

8.1 Introduction 118
8.2 Materials and Methods 120
  8.2.1 Problem Formulation 120
  8.2.2 Research Methodology 121
  8.2.3 Quasi-Anechoic Chamber Description 123
  8.2.4 WIPL-D Simulation Software 124
8.3 Results and Discussion 125
  8.3.1 Test Scenarios and Numerical Simulation Issues 125
  8.3.2 Spatial RCS Characteristics for Model Set A 127
  8.3.3 Spatial RCS Characteristics for Model Set B 133
  8.3.4 Measurements in the Quasi-Anechoic Chamber 134
8.4 Conclusions 138

8.1 INTRODUCTION

Radar Cross-Section (RCS) is a crucial component in modern radar technology and stealth design, serving as a fundamental parameter for evaluating an object's detectability by radar systems. Over the years, numerous studies have been conducted to understand and manipulate RCS, leading to significant advancements in radar systems and air defense applications. There are several well-known approaches for determining the RCS of objects, including both measurement and simulation methods.

The measurement methods are discussed, among others, in [1] and [2]. The main advantage of these methods is their ability to examine real objects, taking into account not only geometric features but also the materials they are made of. The primary disadvantage, however, is the high cost associated with the measurement equipment and the construction of an anechoic chamber. Additionally, for large objects, it may be impossible to ensure anechoic measurement conditions. As a result, measurements are most often conducted in measurement ranges [1], which have limited access. Due to these challenges, measurements are increasingly performed under semi-anechoic [3, 4] or non-anechoic conditions [5]. These methods, however, require measurement equipment that supports time gating, as well as a complex calibration process to eliminate the impact of noise and interference on the results [1–7].

On the other hand, performing measurements under real conditions appears to be a good solution, as it reflects the actual RCS behavior of the tested object. This type of measurement is conducted in [3] for unmanned aerial vehicles (UAVs). UAVs are an interesting area of research due to their growing popularity in reconnaissance and offensive military applications. The accuracy of measurements related to RCS fluctuations is addressed in [8] and [9], where the impact of target RCS fluctuations on radar measurement accuracy is considered.

When operating radar in real conditions, the RCS value fluctuates over time due to factors such as changes in the observation angle as the object moves, as well as the constantly changing environment. This is evidenced by the widespread use of statistical models, with the most popular being the Swerling models. Four different Swerling models of target fluctuation have been defined, and probability density functions are presented for each of the four models. When determining the RCS probability density function based on Swerling models, the mean value of the RCS of the reflecting object is used [10, 11]. Swerling models have also been modified to bring them closer to real-world conditions [12]. However, to accurately determine the scattering properties of a radar target and fit the statistical model of fluctuations to a specific object, it is essential to understand the RCS characteristics as a function of the observation angle. Objects of interest to radars are often characterized by complex geometry, and the RCS values for different observation directions can differ by dozens of dBsm. Knowledge of the object's geometric features and the ability to combine the RCS characteristics with information obtained from radar data can significantly impact the decision-making process and the recognition of the object with which the radar system is dealing.

Since it is not always possible to perform RCS measurements on real objects [13], and the equipment required for such measurements tends to be extensive [14], both direct and asymptotic simulation methods prove to be extremely useful. Comparisons of numerical methods can be found in [15, 16], while comparisons of asymptotic and hybrid methods are presented in [17, 18]. The issue of computational speed in determining RCS, as well as approaches to accelerate the process, has been discussed in [19–21].

According to [22], high-frequency asymptotic methods such as Physical Optics (PO) and Geometrical Optics (GO) are used when the wavelength is much shorter than the dimensions of the target object. However, GO and PO are generally inaccurate near edges or at the boundaries of shaded regions, where they must be supplemented with diffraction and creeping-wave calculations.

In contrast, one of the most widely used full-wave methods is the Method of Moments (MoM), which is considered the most accurate and is often used as a reference for other methods. However, it is less computationally efficient than the Fast Multipole Method (FMM), the Finite-Difference Time-Domain (FDTD) method, the Transmission-Line Matrix (TLM) method, and the Finite Element Method (FEM) [15]. The application of the FDTD method for RCS determination is discussed in [23–25], among others.

It is considered good practice to compare simulation techniques based on differential equations (e.g., FDTD, FEM, and TLM) with those based on integral equations (e.g., MoM), as the fundamental approaches to the solution differ significantly [26]. However, due to the high cost of maintaining licenses for multiple solvers, such comparisons are not always feasible.

High accuracy requires accounting for complex geometry and often renders asymptotic methods inapplicable. On the other hand, full-wave methods such as the MoM demand extremely large computational resources. To mitigate the computational time scaling of MoM, the Multi-Level Fast Multipole Method (MLFMM) is applied in [27], where a range of advancements toward an efficient algorithm for large-scale full-wave monostatic RCS computation is proposed, particularly for structures too large to be handled with MoM.

When determining the RCS of large objects, such as fighter jets, studies are often conducted using scaled models [28, 29]. The study [30] presents simulation and measurement results using a scaled-down F-35 fighter model.

In the research presented in this article, the geometry of the analyzed object is small enough to allow the successful application of MoM. For this purpose, the WIPL-D software, described in Section 8.2, was used.

Another important aspect considered is the influence of construction materials on RCS characteristics. In many cases, assuming that objects are made entirely of metal and modeling them as perfect electric conductors (PEC) introduces a significant simplification. To better approximate real-world conditions in which radar systems operate, we conducted additional simulations that account for construction materials beyond PEC. Issues related to simulation methods incorporating dielectric properties are discussed in [31–33], among others. Meanwhile, RCS measurement-based studies focusing on metasurfaces are presented in [34].

The ability to incorporate material properties into simulations is made possible by the WIPL-D software used in this research. However, ensuring accurate modeling requires precise knowledge of the electromagnetic properties of these materials.

Beyond selecting an appropriate method for determining RCS and accounting for construction materials, this study also considers the accuracy of model geometry. An effort was made to assess the impact of simplifying the airframe of the considered test missile and removing small structural elements such as screws and cable covers. The results of the conducted simulations are presented in Section 8.3. The issue of how 3D model simplifications affect the spatial characteristics of RCS has been explored in [35] and [36], among others.

The accuracy of computational results is influenced by numerous factors. Not only do the choice of calculation method and software tool play a crucial role but also the way the 3D model is prepared—specifically, the level of detail, mesh selection, and consideration of construction materials—should be regarded as essential.

A key novelty of this research is the incorporation of multi-variability in the 3D model, considering different levels of detail, grid element counts, and construction materials. Due to limitations in the number of available software tools, measurement results from a quasi-anechoic chamber were used to validate the determined RCS values. Such research is essential for developing effective 3D object modeling methods for both simulation and measurement, ensuring the desired accuracy while maintaining a reasonable computation time.

8.2 MATERIALS AND METHODS

8.2.1 PROBLEM FORMULATION

Determining the RCS of targets relevant to anti-aircraft systems is a complex and ongoing challenge. The problem lies in accurately determining the spatial characteristics of RCS in a timely manner while ensuring that the results meet the required level of accuracy. Achieving this accuracy depends not only on the measurement or simulation method used but also on how the 3D model of the target object is prepared and the knowledge of the construction materials involved. The complexity arises from the need to balance detailed modeling, computational resources, and the limitations of the available measurement techniques.

8.2.2 RESEARCH METHODOLOGY

The research conducted includes several scenarios described in Section 8.3. These scenarios address both the impact of details on the spatial characteristics of the RCS and construction materials, as well as a comparison of results obtained through simulations and measurements.

The part of the research related to measurements in a quasi-anechoic chamber was carried out to compare the simulation results with those from the measurement setup. Since the presented measurement setup is primarily designed for antenna technology, the results should be considered as indicative.

The WIPL-D software was used to perform the calculations. The simulation required a 3D model of the object under investigation. Since one element of the analysis was to assess the impact of details on the resulting spatial RCS characteristics, the 3D model was prepared in several variants.

The measurement of the RCS is divided into three steps:

1. determining the minimum reflected signal distance,
2. introducing a sample object for reference measurement,
3. measuring the level of the signal reflected from the sample object.

The simulation of the RCS is divided into three steps:

1. preparing geometric models with varying levels of detail,
2. loading geometric models and setting up projects in the software,
3. performing simulations.

The results obtained from the measurements and simulations were analyzed and then compared for the 3D model that, in the authors' opinion, most closely represents the real object for which the measurements were taken.

In order to determine the RCS value, it is necessary to assume that the power in free space is transmitted back to the transmitting/receiving antenna of the system that was used to introduce the radiation. The RCS σ is based on the transition function, which relates the incident and reflected power values according to the relationship [1]:

$$P_R = P_T \frac{G^2 \lambda^2 \sigma}{(4\pi)^3 R^4} \tag{8.1}$$

where P_R is the received power (W), P_T is the transmitted power (W), R is the distance from radar to target (m), G is the antenna gain (dB), λ is the wavelength (m), and σ is the RCS of the aerial target (m^2).

In the first stage of measurements in the quasi-anechoic chamber, the distance in the far field of the antennas was determined. This distance was chosen to achieve the minimum received signal, which was reflected from the walls of the chamber and the elements of the measurement stand, without the metal plate. This setup provides the greatest measurement dynamics in relation to the signal level reflected from the metal plate fixed on the test stand. The distance between the antennas and the test object was set to 2.5 m.

In the second stage, a 30 × 30-cm steel plate was placed on the test bench. The reference measurement was then taken, which involved measuring the signal reflected from the reference plate at each measurement frequency.

In the third stage, the signal reflected from the test object was measured. The relative difference in signal levels, measured in the previous two stages, was used to determine the test object's absorption of electromagnetic (EM) waves. The absorption level was calculated by comparing it to the signal reflected from the steel plate.

Based on the performed three-step measurements, the RCS may be determined according to the following algorithm:

1. Calculation of the theoretical RCS value (for all measurement frequencies) for a flat steel plate, using the following equation:

$$\sigma = \frac{4\pi a^2 b^2}{\lambda^2} \tag{8.2}$$

   where a, b are the width and height of the metal plate, respectively (m) and λ is the wavelength (m).
2. Measuring the incident and reflected power from a flat steel plate:

$$\frac{P_R}{P_T} = \frac{G^2 \lambda^2 \sigma}{(4\pi)^3 R^4} \tag{8.3}$$

3. Determining the value of the proportionality factor K related to the measurement setup (influence of the measurement environment) according to the formula:

$$K = \frac{G^2 \lambda^2}{(4\pi)^3 R^4} \tag{8.4}$$

   The value of the K factor determines the parameters of the measurement system. It depends on the wavelength of the signal and is given in m^{-2}. If a reference object (in the form of a steel plate with dimensions a and b) with a known σ value, determined using Eq. (8.2), is available, the value of the K factor can be determined using the following equation:

$$\frac{P_R}{P_T} = \sigma K \tag{8.5}$$

   The values of P_R and P_T are obtained by measuring the reference object.

4. The measurement of incident and reflected power from the test object as a function of the angle of incidence of the electromagnetic wave, and the determination of σ as a function of the angle of incidence (for specific frequencies), was performed according to Eq. (8.5).

8.2.3 QUASI-ANECHOIC CHAMBER DESCRIPTION

A quasi-anechoic chamber was used to conduct RCS measurements of the real object, corresponding to the geometry of the model in the simulation research.

The main purpose of the quasi-anechoic chamber is to isolate the measurement field from the surrounding environment and ensure electromagnetic silence. It maintains stable environmental conditions (temperature, humidity) during tests, which guarantees the stability of the measuring system parameters and ensures the reliability and repeatability of the measurements.

The measurement system used for the tests was located in two rooms—inside the quasi-anechoic chamber and in the equipment room. The RCS measurement system includes the following instruments: the SMB100A signal generator, FSET22 measurement receiver, DRG118/A horn antennas, Maturo turntable, and quasi-anechoic chamber.

In Fig. 8.1, the symbol E_t represents the electric field strength of the plane wave propagating from the transmitting antenna to the test object. The symbol E_r represents the electric field strength of the plane wave propagating from the test object to the receiving antenna. The symbol E_c represents the electric field coupling between the transmitting and receiving antennas, which is part of the received field by the receiving antenna and affects the dynamic range of the measurement. Note that, for illustrative purposes, the object on the turntable is shown with a sample orientation, which does not correspond to the actual position of the real object during the tests.

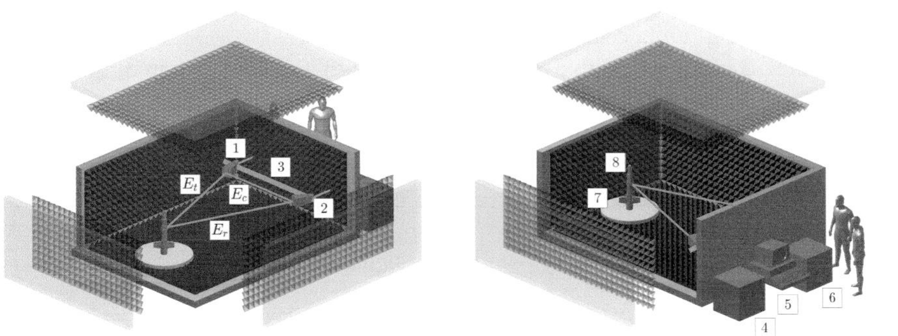

Figure 8.1 General view of the laboratory stand for RCS measurement: 1—movable transmission antenna; 2—movable receive antenna; 3—movable rail; 4—FSET 22 receiver; 5—computer; 6—SMB100A signal generator; 7—Maturo turntable; 8—test object under consideration.

8.2.4 WIPL-D SIMULATION SOFTWARE

WIPL-D is a commercial, general-purpose, full-wave 3D EM solver. It is based on the MoM applied to Surface Integral Equation (MoM/SIE). WIPL-D calculates the unknown equivalent currents across a discretized (meshed) boundary surface.

For accurate and efficient geometric representation of 3D structures, triangles are often the simplest and most versatile choice. Many geometry files, including STL files, use triangles as building blocks. Therefore, MoM/SIE adapted to the triangular mesh seems like the natural choice for triangle-based models. However, MoM/SIE applied to a quadrilateral mesh has been shown to be more efficient [37]. Thus, to run WIPL-D simulations, the triangular mesh must first be transformed into a quadrilateral one. This automated, user-controlled transformation is available as a built-in feature in the WIPL-D software package.

Using a quadrilateral mesh boosts WIPL-D's efficiency compared to other MoM implementations. Solver performance is further enhanced by applying Higher Order Basis Functions (HOBFs) to approximate unknown currents. This enables WIPL-D to solve very large problems, far beyond the reach of conventional MoM codes [38].

As explained, WIPL-D's efficiency results from carefully selecting the best solution for each step in the MoM/SIE procedure, without sacrificing accuracy. For RCS calculations, the high accuracy has been confirmed through comparisons with Austin RCS Benchmarks.

PHYSICAL AND NUMERICAL MODEL OF THE MISSILE

The 9M111 Fagot, a second-generation anti-tank guided missile (North Atlantic Treaty Organization reporting name: AT-4 Spigot), was chosen as the physical model for the numerical simulation of an axisymmetric object. This choice was made for two reasons: the availability of the model and the physical dimensions of the anechoic measurement chamber. Additionally, the decision was made to avoid disclosing the parameters and characteristics of objects currently in service. Thus, a decommissioned missile, not directly related to air combat, was selected.

The missile is 850 mm long, with a diameter of 120 mm and a wingspan of 369 mm. Its mass is about 8 kg. The canards are rotated 45° relative to the wings. For the RCS measurement and modeling process, the airframe coating (including the outer cable sheath) is mainly made of thin aluminum sheet with anti-reflective paint. The control section (behind the rocket engine) and the canards are coated with resole resin, while the tail section is made of steel. The rear wings are constructed from steel sheet containing approximately 0.2% carbon. Standard steel screws are also used.

Based on the physical model, a digital representation of the missile was created in Blender 3D (see Fig. 8.2) and exported in both STL and MATLAB-native MAT file formats. The 3D model has 46,897 vertices and 88,745 faces, maintaining a 1:1 scale. It strikes a balance between accurately mapping the real object and minimizing the computational overhead for WIPL-D simulation software. The model is designed to allow flexible simplifications, such as omitting specific elements or degrading the airframe's shape.

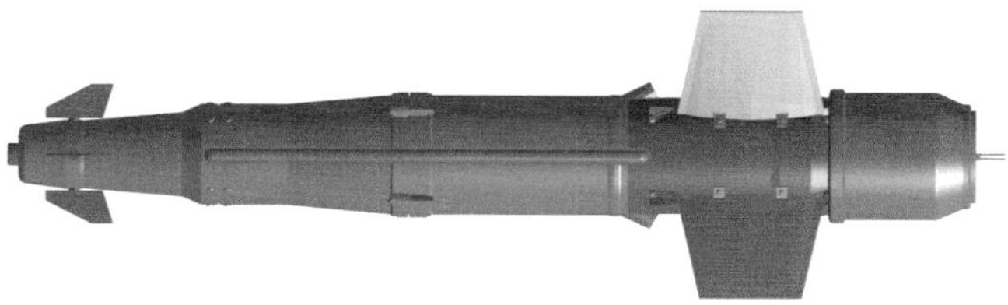

Figure 8.2 9M111 Fagot missile 3D model.

Two sets of simulation models were then prepared in WIPL-D: (a) a simplified set A made entirely of aluminum and (b) a complex set B that incorporates the reflection parameters of the materials listed above (see Figs. 8.3–8.5). A detailed description of the models in sets A and B is shown in Table 8.1. As seen, simplifying the missile's airframe geometry significantly reduces the complexity of the simulation model (see A-1 and A-4 models in Table 8.1).

8.3 RESULTS AND DISCUSSION

8.3.1 TEST SCENARIOS AND NUMERICAL SIMULATION ISSUES

The first scenario involves simulations for a missile made of aluminum (A-labeled models), considering various simplifications of both the airframe and minor components like screws. The second scenario focuses on simulations of a missile with detailed material information, where only minor components are simplified (B-labeled models). Simplification configurations and model IDs are shown in Table 8.1. Simulations were conducted at frequencies of 300 MHz, 1 GHz, and 3 GHz.

The goal of the study was to assess how different simplifications and the knowledge of material parameters affect the spatial characteristics of RCS. The

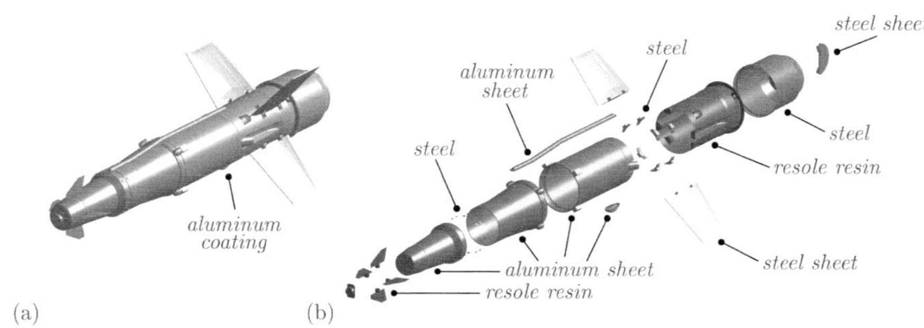

Figure 8.3 Map of materials used for: (a) model set A and (b) model set B.

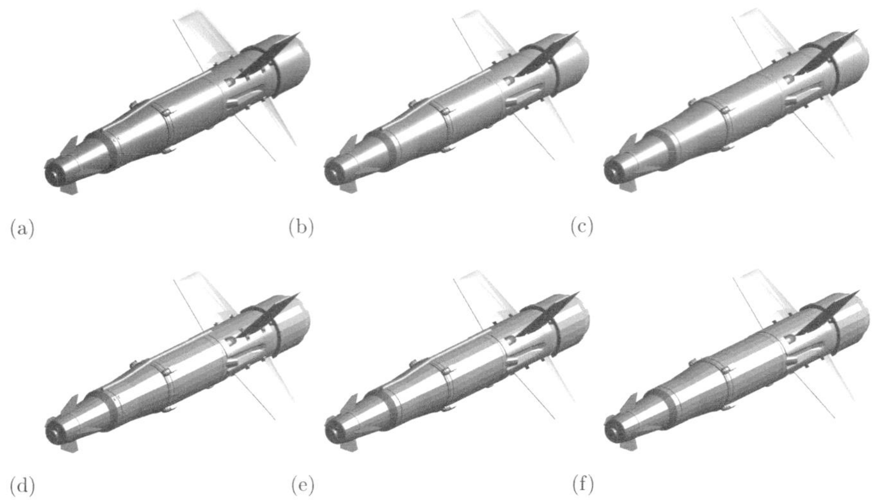

Figure 8.4 Model set A assumed for numerical computations. Each of the missile models (a)–(f) differs in the geometry simplification level (cf. Table 8.1 for detailed description).

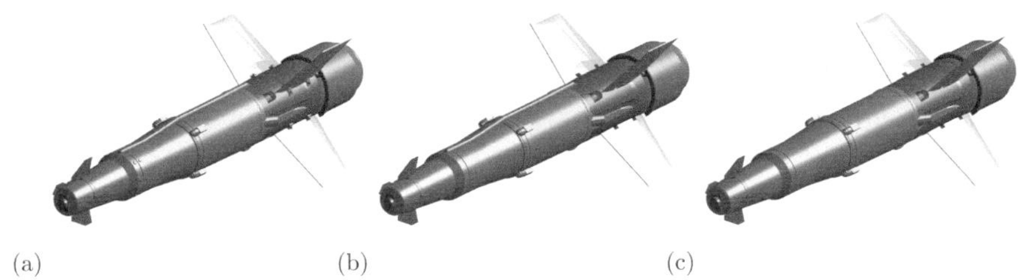

Figure 8.5 Model set B assumed for numerical computations. Similar to the case of set A, the missile models (a)–(c) differ in the geometry simplification level (cf. Table 8.1 for detailed description).

preparation of the 3D geometry is crucial for calculation feasibility and duration. Therefore, studying the impact of different modeling methods is an important part of RCS research for air and missile defense targets.

All model sets were initially prepared as STL files in an external environment. They were then loaded into the WIPL-D Pro environment and processed following the same methodology, with identical mesh properties, polarization, and angular resolution. All simulations used the full-wave 3D EM WIPL-D solver based on MoM/SIE, as described in Section 8.2.

The third scenario compares simulation results at 3 GHz for models without structural simplifications (A-1 and B-1) and the results from measurements of a real object in a quasi-anechoic chamber. The tests were performed in a quasi-anechoic chamber, designed for electromagnetic compatibility testing and antenna technology, as

Table 8.1
Simulation Model Set Description

Model set	ID	Material Map	Vertices	Faces	Reference	Simplifications
	A-1	100% aluminum	46,897	88,745	Fig. 8.4a	none
	A-2	100% aluminum	45,441	86,249	Fig. 8.4b	no screws
	A-3	100% aluminum	45,233	85,859	Fig. 8.4c	(1) no screws, (2) no cable cover
A	A-4	100% aluminum	6,158	12,086	Fig. 8.4d	airframe simplified
	A-5	100% aluminum	4,702	9,590	Fig. 8.4e	(1) airframe simplified, (2) no screws
	A-6	100% aluminum	4,494	9,200	Fig. 8.4f	(1) airframe simplified, (2) no screws, (3) no cable cover
	B-1	cf. Fig. 8.3b	46,897	88,745	Fig. 8.5a	none
B	B-2	cf. Fig. 8.3b	45,441	86,249	Fig. 8.5b	no screws
	B-3	cf. Fig. 8.3b	45,233	85,859	Fig. 8.5c	(1) no screws, (2) no cable cover

explained in Section 8.2. Since the chamber is not dedicated to RCS testing, the measurements should be considered as a verification of the overall RCS level and characteristic shape from the simulations.

Simulation and measurement data were compared using the MATLAB environment.

8.3.2 SPATIAL RCS CHARACTERISTICS FOR MODEL SET A

Simulation results for spatial RCS characteristics at 300 MHz are shown in Figs. 8.6 and 8.7 for the most accurate model (A-1) and the most simplified model (A-6). Comparing these characteristics, no significant differences are noticeable. As expected, the spatial RCS characteristics have a smooth surface. Introducing simplifications to the 3D geometry of the model has little effect. Neither airframe simplifications nor the removal of screws or cable covers significantly affect the spatial RCS characteristics.

Only cuts of the characteristics reveal changes caused by the simplifications (see Fig. 8.8). However, evaluating differences based on selected intersection profiles does not provide complete information about the directions of the largest differences between the spatial RCS characteristics.

To illustrate the differences in spatial RCS characteristics for different model simplifications, the similarity of RCS characteristics is shown in Fig. 8.9. A color map is used to highlight positive (red) and negative (blue) values. This clearly shows the directions where the RCS has changed due to modifications in the missile's 3D geometry. At a frequency of 300 MHz, the absolute similarity values of the RCS characteristics do not exceed 0.5 dBsm.

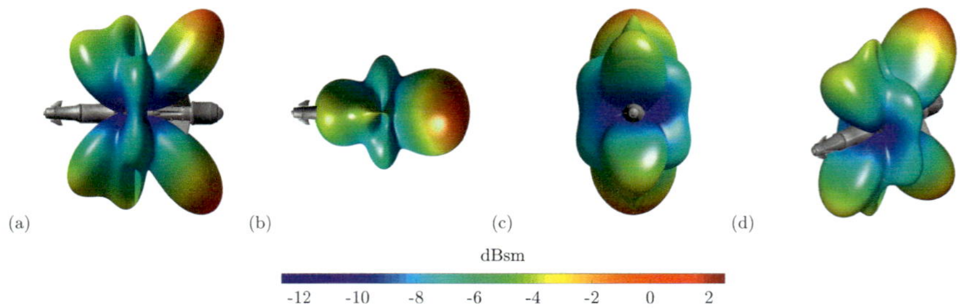

Figure 8.6 Spatial RCS characteristics of model A-1 (most complex) obtained at 300 MHz: a) side view; (b) top view; (c) front view; and (d) isometric view.

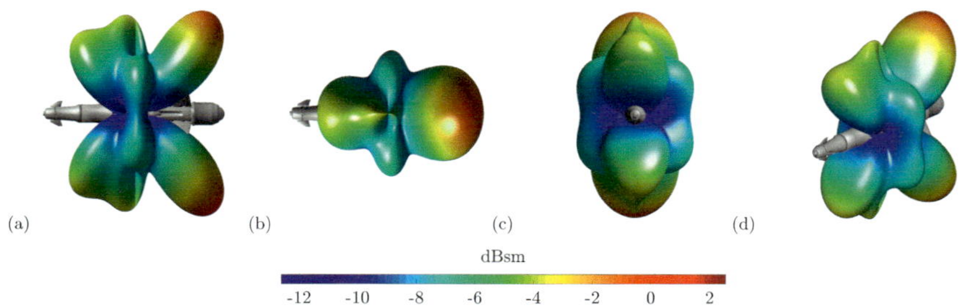

Figure 8.7 Spatial RCS characteristics of model A-6 (most simplified) obtained at 300 MHz: (a) side view; (b) top view; (c) front view; and (d) isometric view.

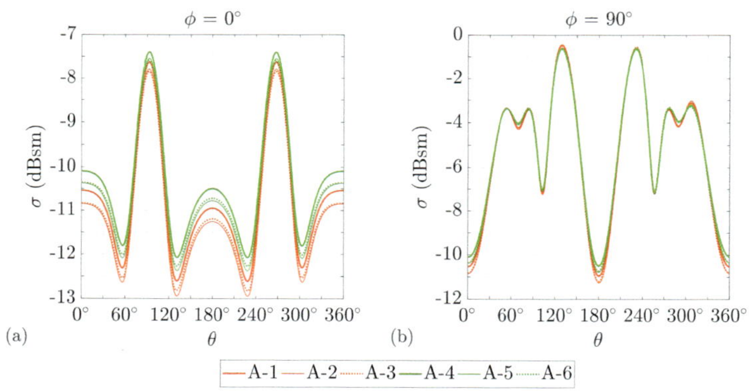

Figure 8.8 RCS envelopes obtained at 300 MHz for the intersection profiles relative to the longitudinal axis of the airframe for: (a) top view and (b) side view of the missile.

It is clear that, for very high frequency (VHF) band frequencies, simplifying the model by removing details like screws, small hooks, cable covers, or airframe elements does not significantly affect the results. This is especially beneficial for simulations of large objects, as it positively impacts computation time.

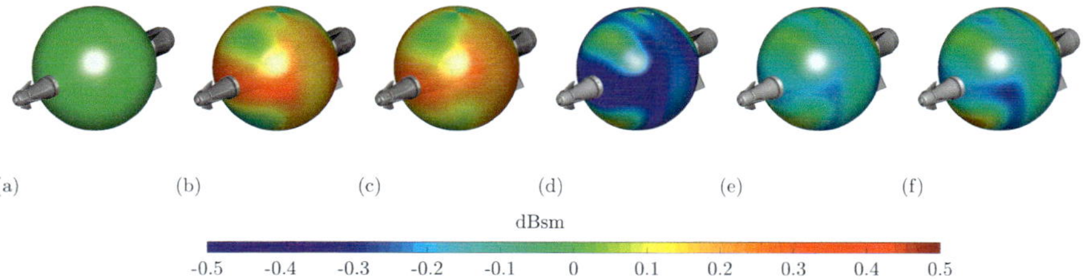

Figure 8.9 The similarity of RCS spatial characteristics obtained at 300 MHz to those of the A-1 model determined for: (a) A-1 (perfect similarity); (b) A-2; (c) A-3; (d) A-4; (e) A-5; and (f) A-6 missile model.

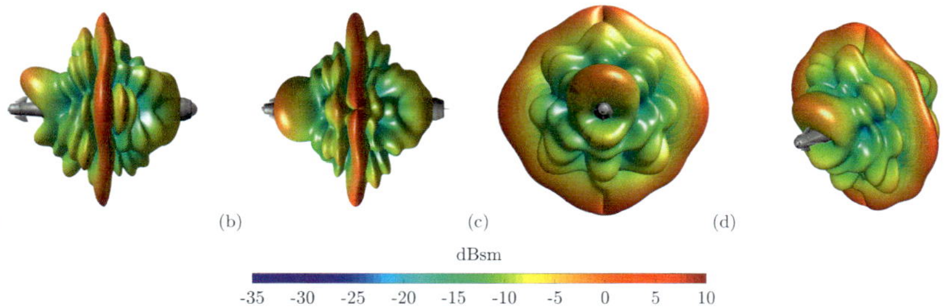

Figure 8.10 Spatial RCS characteristics of model A-1 (most complex) obtained at 1 GHz: (a) side view; (b) top view; (c) front view; and (d) isometric view.

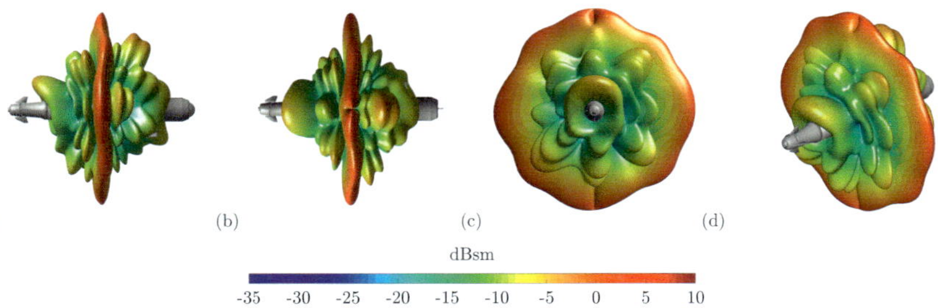

Figure 8.11 Spatial RCS characteristics of model A-6 (most simplified) obtained at 1 GHz: (a) side view; (b) top view; (c) front view; and (d) isometric view.

Simulation results at 1 GHz for model set A, comparing the most accurate and most simplified models, are shown in Figs. 8.10 and 8.11. Differences are noticeable in certain directions of the spatial RCS characteristics, particularly where local maxima occur.

The envelopes shown in Figs. 8.12 and 8.13 provide important information. In the longitudinal cross-section (top view, Fig. 8.12), differences are visible in the front part of the spatial RCS characteristics (θ values close to 0° and 360°).

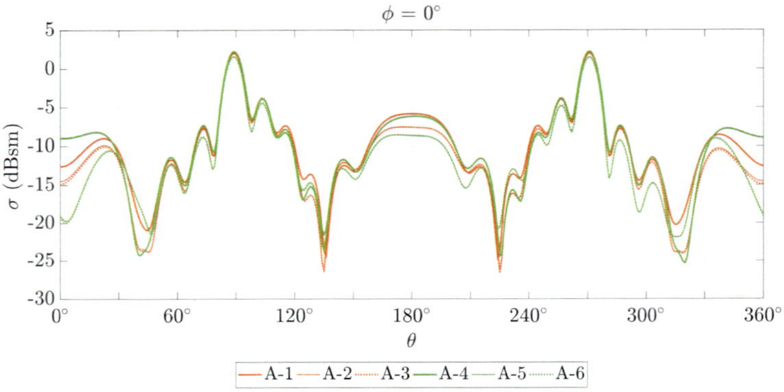

Figure 8.12 RCS envelopes obtained at 1 GHz for the intersection profiles relative to the longitudinal axis of the airframe—top view of the missile.

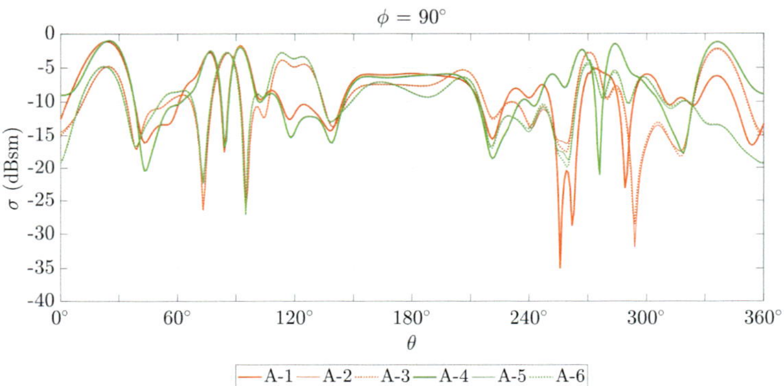

Figure 8.13 RCS envelopes obtained at 1 GHz for the intersection profiles relative to the longitudinal axis of the airframe—side view of the missile.

The changes are much more noticeable in the longitudinal section for the side view (Fig. 8.13). It is also worth noting that airframe simplifications have a greater impact on the shape of the RCS characteristics than the removal of details. The exception is the RCS values for θ in the range from about 260° to 320°, where significant differences appear between the longitudinal cross-sections (side view of the missile, Fig. 8.13) for the A-1 and A-4 models. In the remaining range of θ values, the RCS values for this longitudinal cross-section are most similar across all the models in set A.

In Fig. 8.14, which shows the similarity of RCS spatial characteristics, the influence of successive simplifications is clearly evident. The analysis of the longitudinal cross-sections confirms that changes in RCS characteristics are much greater after removing details than after simplifying the airframe geometry. In some directions, the differences in RCS values reach up to 25 dBsm.

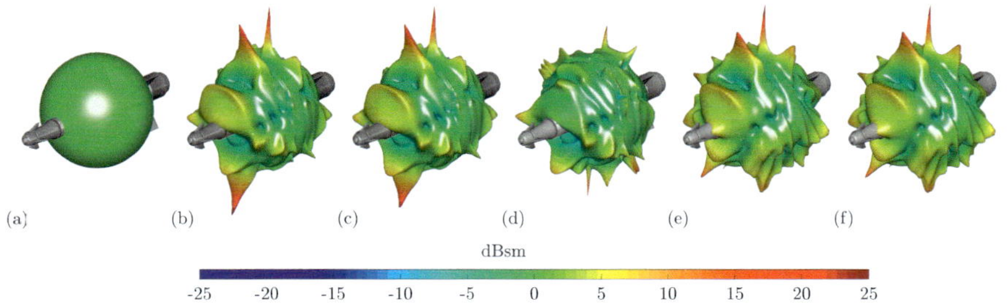

Figure 8.14 The similarity of RCS spatial characteristics obtained at 1 GHz to those of the A-1 model determined for: (a) A-1 (perfect similarity); (b) A-2; (c) A-3; (d) A-4; (e) A-5; and (f) A-6 missile model.

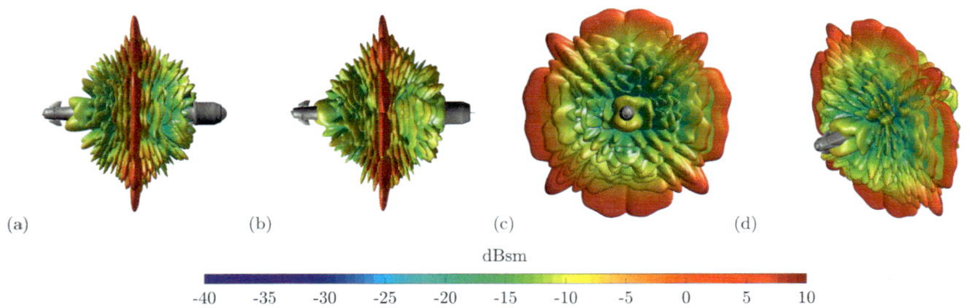

Figure 8.15 Spatial RCS characteristics of model A-1 (most complex) obtained at 3 GHz: (a) side view; (b) top view; (c) front view; and (d) isometric view.

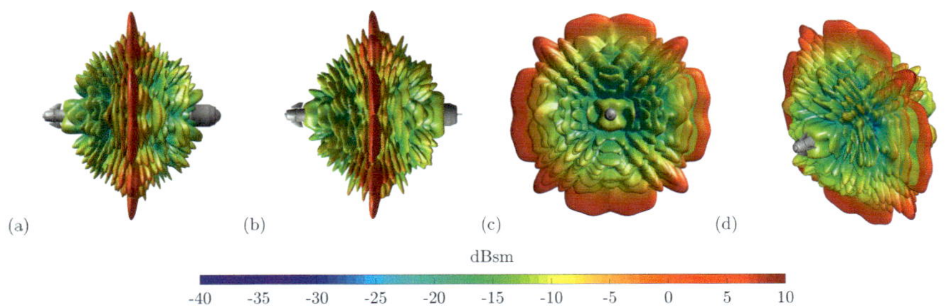

Figure 8.16 Spatial RCS characteristics of model A-6 (most simplified) obtained at 3 GHz: (a) side view; (b) top view; (c) front view; and (d) isometric view.

Spatial RCS characteristics simulated at 3 GHz for the most accurate model (A-1) and the most simplified model (A-6) are shown in Figs. 8.15 and 8.16, respectively. As with the RCS characteristics at 300 MHz, the differences between the models in set A are not easily noticeable. This is due to the high dynamics of changes in the RCS characteristics as the direction of the incident wave changes, which is further influenced by the increase in frequency.

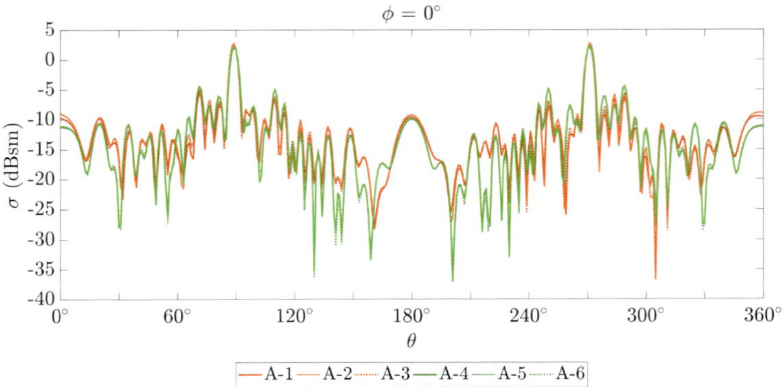

Figure 8.17 RCS envelopes obtained at 3 GHz for the intersection profiles relative to the longitudinal axis of the airframe—top view of the missile.

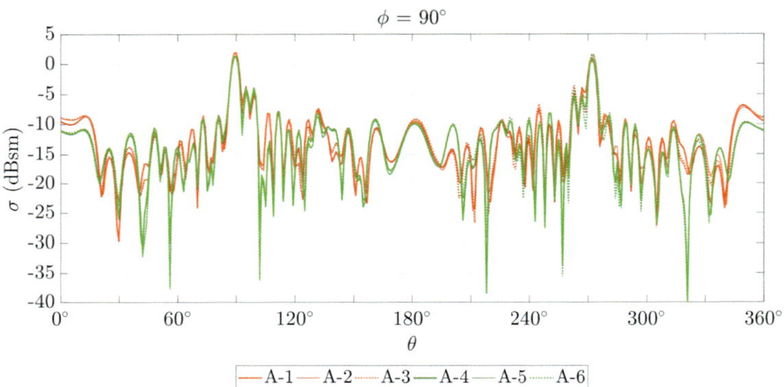

Figure 8.18 RCS envelopes obtained at 3 GHz for the intersection profiles relative to the longitudinal axis of the airframe—side view of the missile.

Analyzing the RCS values in Figs. 8.17 and 8.18, which show longitudinal cross-sections of the RCS spatial characteristics for model set A, we can see that the largest differences occur at local minima. The discrepancies between the cross-sections for models A-1 and A-4 are clearly visible. This indicates that airframe simplification has a significant impact on the spatial RCS characteristics, while removing details from the 3D model does not cause such substantial changes.

Again, the visualizations of the similarity of spatial RCS characteristics, shown in Fig. 8.19, were helpful in analyzing the changes caused by applying simplifications to the missile's 3D geometry. For models A-2 and A-3, where screws and the cable cover were removed, the changes are mild and much smaller than when these details are kept, along with airframe simplifications in model A-4. Removing details along with the simplified airframe, similar to the most accurate version, does not cause significant changes in the spatial RCS characteristics.

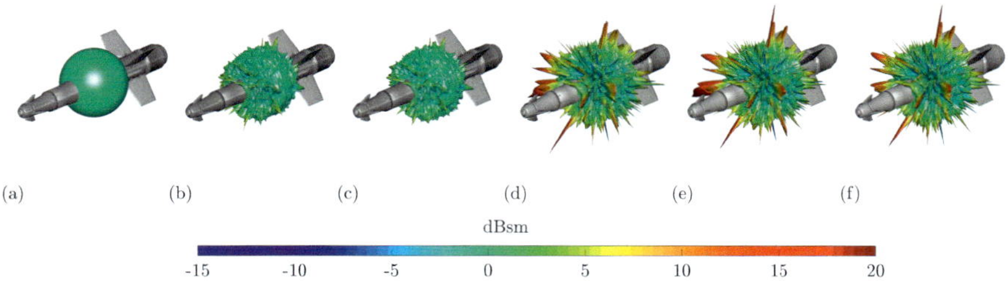

Figure 8.19 The similarity of RCS spatial characteristics obtained at 3 GHz to those of the A-1 model determined for: (a) A-1 (perfect similarity); (b) A-2; (c) A-3; (d) A-4; (e) A-5; and (f) A-6 missile model.

To summarize the observations for the model made from a single construction material (aluminum), it can be stated that, as expected for lower frequencies (in the VHF band), both airframe simplifications and the removal of details from the 3D geometry do not significantly affect the spatial RCS characteristics. This means the 3D geometry can be greatly simplified and have fewer details.

However, as the frequency increases, the behavior of the spatial characteristics becomes harder to predict. At 1 GHz, the removal of details leads to noticeable changes in the spatial RCS characteristics, while airframe simplifications can be applied successfully without significantly altering its shape.

Interestingly, at 3 GHz, the effect of the simplifications changes. Simplifying the airframe causes significant changes, while removing details has a minimal effect on the spatial RCS characteristics.

These observations are crucial, as they highlight situations where small details have a greater impact on the spatial RCS characteristics than simplifying the main structural element of the object.

8.3.3 SPATIAL RCS CHARACTERISTICS FOR MODEL SET B

The next scenario considered involved determining the spatial RCS characteristics for model set B of the missile, taking into account its structural materials (see Fig. 8.5). This set only included simplifications related to detail removal, with no airframe simplification.

It is important to note that the model preparation process differed slightly from that of the fully aluminum model. In this case, the structural elements of the model needed to be loaded sequentially, with their electromagnetic properties defined independently. As a result, even though the mesh algorithm settings remained the same, the final mesh for models in set B is slightly different from that of set A, which introduces some differences in the spatial characteristics between sets A and B.

Figs. 8.20 and 8.21 show the spatial RCS characteristics for models B-1 (the most complex) and B-3 (the most simplified) at a 3 GHz frequency.

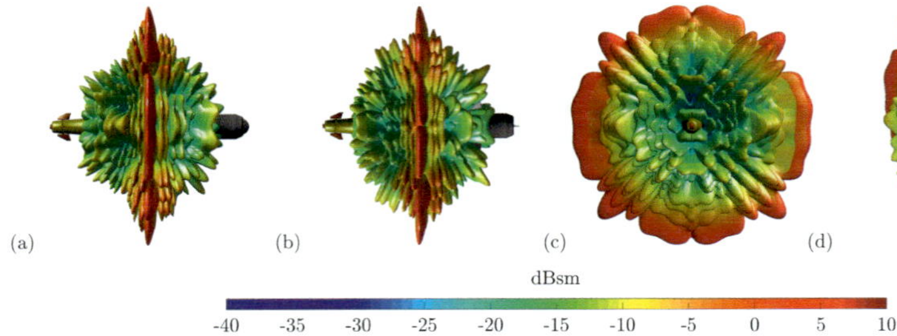

Figure 8.20 Spatial RCS characteristics of model B-1 (most complex) obtained at 3 GHz: (a) side view; (b) top view; (c) front view; and (d) isometric view.

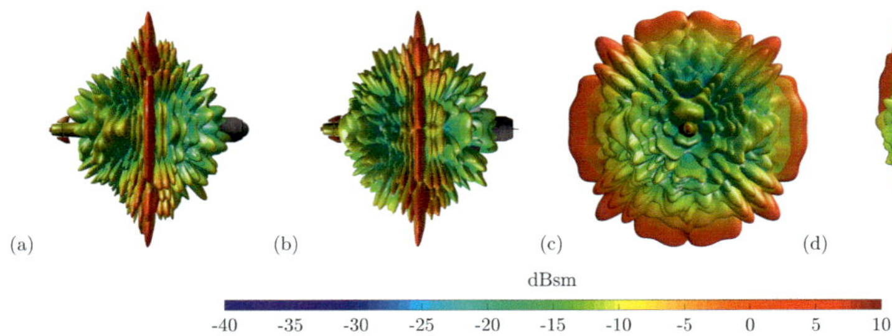

Figure 8.21 Spatial RCS characteristics of model B-3 (most simplified) obtained at 3 GHz: (a) side view; (b) top view; (c) front view; and (d) isometric view.

As with the aluminum model (set A), the RCS values at 3 GHz change dynamically with the direction of the incident wave, making it difficult to assess the impact of changes made to the 3D geometry.

The similarity of the spatial RCS characteristics for the B-model sets is shown in Fig. 8.22. Several directions of the incident wave can be observed for which the differences in RCS values with respect to the B-1 model reach about 20 dBsm.

The RCS envelopes for the intersection profiles relative to the longitudinal axis for model set B shown in Figs. 8.23 and 8.24 are referenced to the most accurate model from model set A (A-1). Comparing the envelopes, it can be seen that greater differences are apparent between the RCS spatial characteristics envelopes of models A-1 and B-1 than between the models in set B. It follows that the inclusion of structural materials brings greater changes in RCS characteristics than its simplification of the model by details.

8.3.4 MEASUREMENTS IN THE QUASI-ANECHOIC CHAMBER

The final stage of the research involved measurements in a quasi-anechoic chamber at a frequency of 3 GHz. The tests were carried out in the horizontal plane with a resolution of 1° and vertical polarization to compare with the simulation results for models A-1 and B-1. Measurements were carried out on the laboratory setup

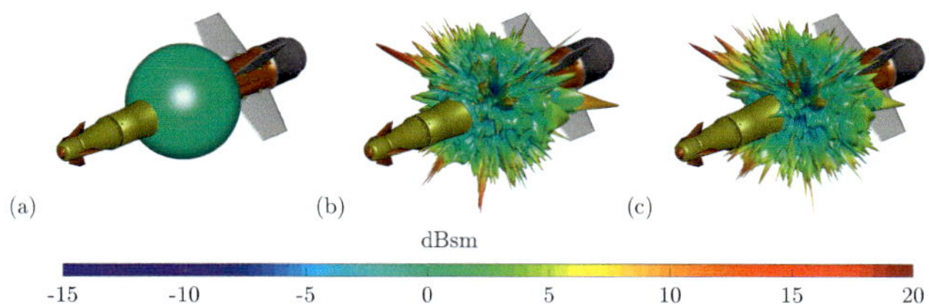

Figure 8.22 The similarity of RCS spatial characteristics obtained at 3 GHz to those of the B-1 model determined for: (a) B-1 (perfect similarity); (b) B-2; and (c) B-3 missile model.

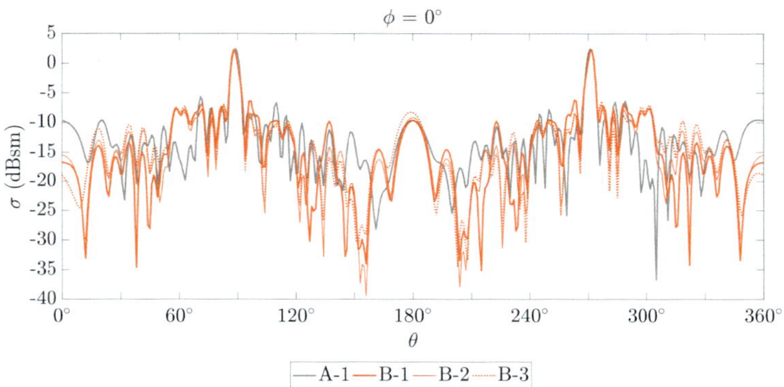

Figure 8.23 RCS envelopes obtained at 3 GHz for the intersection profiles relative to the longitudinal axis of the airframe—top view of the missile.

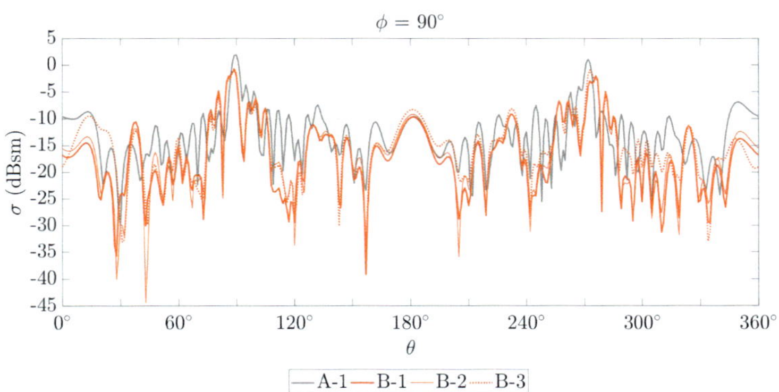

Figure 8.24 RCS envelopes obtained at 3 GHz for the intersection profiles relative to the longitudinal axis of the airframe—side view of the missile.

described in Section 8.2, using the objects shown in Fig. 8.25. Simulations were performed in the WIPL-D environment.

The first measurement was taken for a reference object, a 30 × 30-cm steel plate. The results of the measurements and calculations are shown in Fig. 8.26. The

(a) (b)

Figure 8.25 Experimental measurements in a quasi-anechoic chamber: (a) 30 × 30-cm steel plate for calibrating the laboratory stand and (b) 9M111 Fagot missile under test.

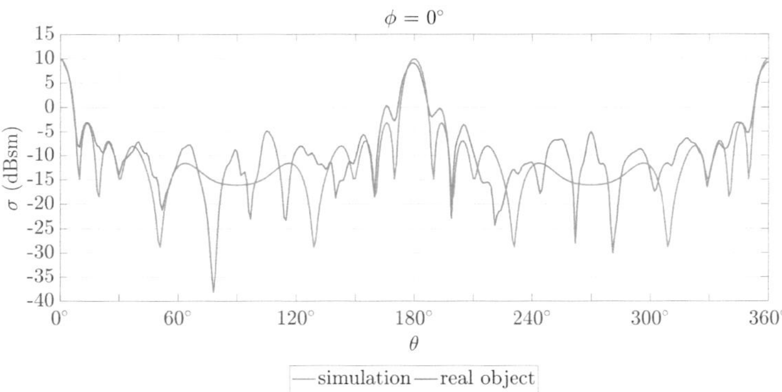

Figure 8.26 Comparison of the reference object measurement result with computer simulations: RCS envelopes obtained at 3 GHz for the intersection profiles relative to the longitudinal axis of the plate—top view of the plate.

greatest agreement between measurement and simulation results was observed when the plate was illuminated from the front and rear. In contrast, significant discrepancies appeared when the plate was illuminated from the sides. The root mean square error (RMSE) for the angle range of (–5°, 5°) was 0.46 dBsm, while for (–10°, 10°) it was 2.08 dBsm. Further increasing the angular range caused a significant rise in RMSE, reaching 3.6 dBsm for angles of (–30°, 30°). However, since only one measurement was made, the results should be treated as indicative rather than definitive. It is clear that for the wave incident on a front flat surface of test plate, both the shape of the measured and simulated characteristics and the obtained RCS values are similar. The differences mainly lie in the deeper minima for angles corresponding to illumination of the plate from the front and a smoother shape when the plate is illuminated from the side.

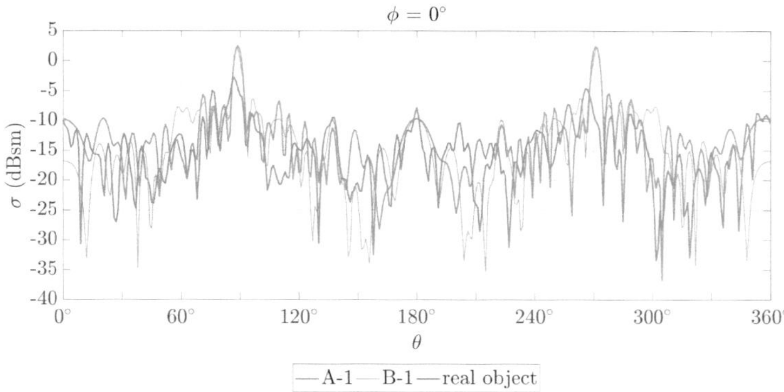

Figure 8.27 Comparison of the real object measurement result with computer simulations: RCS envelopes obtained at 3 GHz for the intersection profiles relative to the longitudinal axis of the airframe—top view of the missile.

The target object used for measurements differs slightly from model B-1 (cf. Fig. 8.3b), particularly in the front part. The parts modeled from resol resin are made of different materials in the real missile. The forward fins are made of metalized material, treated as aluminum, while the fuse and the rest of the fuselage are covered with paint. This object was previously used for educational purposes, and its design was not intended for measurements in a quasi-anechoic chamber. However, due to the availability of detailed information about the geometry and construction materials, it was decided to use it in this research. Since it was not possible to investigate the electromagnetic properties of these materials without damaging the structure, measurements were taken despite the discrepancies between the B-1 model and the actual object.

The data obtained from the chamber measurements were compared with the simulation data for models A-1 and B-1. It should be noted that the values from both simulations and measurements are similar, with both the shape of the characteristics and the RCS level being preserved.

In Fig. 8.27, it can be seen that measurements from the front of the object show better agreement with the simulation for the A-1 model, which assumes a single construction material (aluminum). This is due to the covering of parts that should be made of resole resin with other materials, such as aluminum-like tape and paint. On the other hand, for the measurements taken from the rear of the object, better alignment with the B-1 model simulation results is observed. This is because the materials of the real object match well with the B-1 model. However, a subtle shift in the measured results can be seen, likely caused by imperfections in the measurement setup.

Another factor that could affect the similarity between the measurements and simulations is that, for the reference object with a maximum dimension of 30 cm, the far zone condition $R_{\min} \geq 2D^2/\lambda$ [1] was satisfied, while for the target object with a maximum dimension of 85 cm, it was not. To check how including near-zone simulations would affect the match between measurements and simulations, additional calculations were made at a distance of 2.5 m. However, since these calculations did not result in significant changes or improve the compatibility between the simulation and measurement results, only the far-zone simulation results, as previously discussed, were included.

For the target object, as with the reference object, only one measurement was performed. Therefore, the measured RCS values should be regarded as confirmation of the correctness of the simulation results in terms of order of magnitude and general agreement with the RCS contour.

8.4 CONCLUSIONS

Determining the RCS for targets relevant to anti-aircraft and anti-missile systems is a complex task. It requires a thorough understanding of the 3D geometry of the targets, along with careful preparation of models tailored to the capabilities of the simulation software and EM solvers used. Additionally, the available computational resources and desired computation time play a key role. Simplifying the geometry of the model for simulation purposes is also necessary to achieve accurate results within a reasonable time frame, without causing excessive computational overhead.

The results from Section 8.3 highlight the impact of simplifying 3D models on the spatial RCS characteristics at different frequencies. For the VHF band, simplifying the model by removing certain details—such as small structural elements—does not significantly affect the overall RCS. However, for higher frequencies, such as 1 GHz and 3 GHz, the removal of details like screws or cable covers has a greater impact, especially at 1 GHz, where small elements influence the RCS more than the airframe simplifications. On the other hand, simplifying the geometry at 3 GHz proved more beneficial to the spatial RCS characteristics.

Another significant finding of this study is the importance of accurately modeling the electromagnetic properties of construction materials. The results in Section 8.2 show that the electromagnetic properties of materials have a much greater effect on the resulting spatial RCS than simplifications to the geometry itself. This emphasizes the importance of incorporating accurate material properties into simulation models, as these properties significantly influence the RCS predictions.

Additionally, the coatings and finishes applied to the materials are also critical. The comparison between simulation and measurement results demonstrated that knowledge of both the materials and their coatings—such as the use of aluminum-like tape and paint—can significantly affect the resulting RCS. For example, the use of such coatings in the measurements taken from the front of the object led to a better match with the A-1 model simulation, which assumed aluminum as the sole material.

The experimental measurements and simulations underscore the importance of having precise knowledge of the materials and coatings of targets. This knowledge is crucial not only for the accuracy of the RCS model but also for its implications in real-world air and missile defense systems. Understanding the electromagnetic properties of materials and coatings can directly influence the detectability and recognition of objects, which is crucial for defense applications.

Future research will focus on expanding the study to different types of aerial targets, incorporating various levels of model accuracy, and comparing results from different numerical methods, such as MoM and FDTD. As in this study, future work will continue to validate simulation results with experimental data, ensuring the reliability of the models for practical applications.

AUTHOR CONTRIBUTIONS

Conceptualization, W.B. and M.W.; methodology, W.B., M.W., A.K., R.P., and M.B.; software, B.N., M.T., B.K., K.K., and W.B.; validation, W.B., M.W., B.N., and L.N.; formal analysis, W.B., M.W., K.K., and A.K.; investigation, W.B., M.W., R.P., M.B., and B.N.; resources, W.B., M.W. and A.K.; data curation, W.B., M.W., B.N., M.T., B.K., R.P., and M.B.; writing—original draft preparation, W.B. and M.W.; writing—review and editing, W.B., M.W., K.K., and A.K.; visualization, W.B.; supervision, A.K.; project administration, W.B.; funding acquisition, W.B. All authors have read and agreed to the published version of the manuscript.

DATA AVAILABILITY STATEMENT

The data that support the findings of this chapter are available from the corresponding author, M.W., upon reasonable request.

FUNDING

This work was financed by the Military University of Technology under the research project UGB 22-825/2023.

REFERENCES

1. IEEE Recommended Practice for Radar Cross-Section Test Procedures. 2020. *IEEE Std 1502-2020 (Revision of IEEE Std 1502-2007)*:1–78.
2. Miacci, M., and M. Rezende. 2012. Basics on radar cross section reduction measurements of simple and complex targets using microwave absorbers. In: *Applied Measurement Systems*, ed. Z. Haq, 361–389. London: IntechOpen Ltd.
3. Leye, P.O., D. Martinez, S. Aldhaheri, et al. Stepped-Frequency CW RCS Measurement in Semi-Anechoic Chamber. In: Proc. of the 2022 Antenna Measurement Techniques Association Symposium (AMTA). Denver (CO, USA), October 09–14, 2022, pp. 1–4. https://doi.org/10.23919/AMTA55213.2022.9954995.
4. Rosamilia, M., A. Aubry, A. Balleri, et al. RCS Measurements of UAVs and Their Statistical Analysis. In: Proc. of the 2022 IEEE 9th International Workshop on Metrology for AeroSpace (MetroAeroSpace). Pisa (Italy), June 27–29, 2022, pp. 179–184. https://doi.org/10.1109/MetroAeroSpace54187.2022.9856394.
5. Mihai, I.V., A. Constantin, S. Bucuci, et al. 2022. A single-antenna method and post-processing strategy for radar cross-section measurements at near-field ranges. *Sensors* 22(19):7453.
6. Sundermeier, M.F., and D. Fischer. 2021. Compact radar cross-section measurement setup and performance evaluation *Advances in Radio Science* 19:147–152.

7. Soltane, A., G. Andrieu, and A. Reineix. Monostatic Radar Cross-Section Estimation of Canonical Targets in Reverberating Room Using Time-Gating Technique. In: Proc. of the 2018 International Symposium on Electromagnetic Compatibility (EMC EUROPE). Amsterdam (Netherlands), August 27–30, 2018, pp. 355–359. https://doi.org/10.1109/EMCEurope.2018.8485084.
8. Swerling, P., and W.L. Peterman. 1990. Impact of target RCS fluctuations on radar measurement accuracy. *IEEE Transactions on Aerospace and Electronic Systems* 26(4):685–686.
9. Lank, G.W. 1991. Radar measurement accuracy for fluctuating targets. *IEEE Transactions on Aerospace and Electronic Systems* 27(6):868–870.
10. Swerling, P. 1960. Probability of detection for fluctuating targets. *IRE Transactions on Information Theory* 6(2):269–308.
11. Shin, S.J. 2017. Radar measurement accuracy associated with target RCS fluctuation. *Electronics Letters* 53(11):750–752.
12. Backes, T., and L.D. Smith. Improved RCS Model for Censored Swerling III and IV Target Models. In: Proc. of the 2013 IEEE Aerospace Conference. Big Sky (MT, USA), March 02–09, 2013, pp. 1–4. https://doi.org/10.1109/AERO.2013.6497433.
13. Tice, T.E. 1990. An overview of radar cross section measurement techniques. *IEEE Transactions on Instrumentation and Measurement* 39(1):205–207.
14. Blacksmith, P, R.E. Hiatt, and R.B. Mack. 1965. Introduction to radar cross-section measurements. *Proceedings of the IEEE* 53(8):901–920.
15. Sahoo, N.K., R.K. Parida, and D.C. Panda. Comparison of Numerical Methods in RCS Computation of Corner Reflectors. In: Proc. of the 2018 International Conference on Applied Electromagnetics, Signal Processing and Communication (AESPC). Bhubaneswar (India), October 22–24, 2018, pp. 1–3. https://doi.org/10.1109/AESPC44649.2018.9033445.
16. Yu, D., and M. Zhang. Comparison of Various Full-Wave Methods in Calculating the RCS of Inlet. In: Proc. of the 2008 International Conference on Microwave and Millimeter Wave Technology. Nanjing (China), April 21–24, 2008, pp. 1018–1021. https://doi.org/10.1109/ICMMT.2008.4540583.
17. Bouche, D.P., F.A. Molinet, and R. Mittra. 1993. Asymptotic and hybrid techniques for electromagnetic scattering. *Proceedings of the IEEE* 81(12):1658–1684.
18. Tasić, M.S., and B.M. Kolundžija. 2011. Efficient analysis of large scatterers by physical optics driven method of moments. *IEEE Transactions on Antennas and Propagation* 59(8):2905–2915.
19. Reddy, C.J., M.D. Deshpande, C.R. Cockrell, et al. 1998. Fast RCS computation over a frequency band using method of moments in conjunction with asymptotic waveform evaluation technique. *IEEE Transactions on Antennas and Propagation* 46(8):1229–1233.
20. Wang, Z., and W. Nie. Fast Calculation of Wideband RCS of Objects by Combining Improved Ultra-wide Band Characteristic Basis Function Method and AWE Technique. In: Proc. of the 2016 Progress in Electromagnetic Research Symposium (PIERS). Shanghai (China), August 08–11, 2016, pp. 1156–1159. https://doi.org/10.1109/PIERS.2016.7734605.
21. Li, X., Z. Dong, X. Chen, et al. Accelerated Iterative Solution of Monostatic RCS from PEC Targets via Randomized Algorithm. In: Proc. of the 2019 International Applied Computational Electromagnetics Society

Symposium—China (ACES). Nanjing (China), August 08–11, 2019, pp. 1–2. https://doi.org/10.23919/ACES48530.2019.9060546.
22. Uluisik, C., G. Cakir, M. Cakir, et al. 2008. Radar cross section (RCS) modeling and simulation, part 1: a tutorial review of definitions, strategies, and canonical examples. *IEEE Antennas and Propagation Magazine* 50(1):115–126.
23. Xiaogang, X., L. Wei, L. Ying, et al. Using LU Decomposition in FDFD for Fast Calculation of Monostatic RCS. In: Proc. of the 2014 7th International Conference on Intelligent Computation Technology and Automation. Changsha (China), October 25–26, 2014, pp. 887–889. https://doi.org/10.1109/ICICTA.2014.212.
24. Sahoo, N.K., R.K. Parida, and D.C. Panda. A Case Study on using CPML and PML Boundary Conditions in FDTD for RCS Calculation. In: Proc. of the 2018 International Conference on Applied Electromagnetics, Signal Processing and Communication (AESPC). Bhubaneswar (India), October 22–24, 2018, pp. 1–3. https://doi.org/10.1109/AESPC44649.2018.9033411.
25. Cakir, Go., M. Cakir, and L. Sevgi. 2008. Radar cross section (RCS) modeling and simulation, Part 2: A novel FDTD-based RCS prediction virtual tool for the resonance regime. *IEEE Antennas and Propagation Magazine* 50(2):81–94.
26. IEEE Draft Recommended Practice for Validation of Computational Electromagnetics Computer Modeling and Simulations. 2010. IEEE P1597.2/D7.2, pp. 1–117.
27. Borries, O., E. Jorgensen, and P. Meincke. Monostatic RCS of Electrically Large Structures Using Higher-order MLFMM. In: Proc. of the 2017 11th European Conference on Antennas and Propagation (EUCAP). Paris (France), March 19–24, 2017, pp. 872–876. https://doi.org/10.23919/EuCAP.2017.7928658.
28. Kong, M., M.S. Chen, X.Y. Cao, et al. Fast Calculation of Monostatic RCS by Calderón Multiplicative Preconditioner and Compressive Sensing. In: Proc. of the 2018 12th International Symposium on Antennas, Propagation and EM Theory (ISAPE). Hangzhou (China), December 03–06, 2018, pp. 1–3. https://doi.org/10.1109/ISAPE.2018.8634229.
29. Yamada, Y., N. Michishita, and Q.D. Nguyen. Calculation and Measurement Methods for RCS of a Scale Model Airplane. In: Proc. of the 2014 International Conference on Advanced Technologies for Communications (ATC 2014). Hanoi (Vietnam), October 15–17, 2014, pp. 69–72. https://doi.org/10.1109/ATC.2014.7043358.
30. Herda, D.L., J. Suryana, and A. Izzuddin. Radar Cross Section of F35: Simulation and Measurement. In: Proc. of the 2020 6th International Conference on Wireless and Telematics (ICWT). Yogyakarta (Indonesia), September 03–04, 2020, pp. 1–6. https://doi.org/10.1109/ICWT50448.2020.9243627.
31. Im, H.R., W. Kim, Y.H. Noh, et al. 2022. RCS estimation of singly curved dielectric shell structure with PMCHWT method and experimental verification. *Sensors* 22(3):734.
32. Mohammadzadeh, H., A.Z. Nezhad, and Z.H. Firouzeh. Modified Physical Optics Approximation for RCS Calculation of Dielectric Coated PEC with Axial Symmetry. In: Proc. of the 2013 21st Iranian Conference on Electrical Engineering (ICEE). Mashhad (Iran), May 14–16, 2013, pp. 1–5. https://doi.org/10.1109/IranianCEE.2013.6599811.
33. Seker, S.S., and G. Apaydin. Electromagnetic Scattering Properties of Thin Curved Dielectric Surface and Cylinder. In: Proc. of the 2007 Asia-Pacific Microwave Conference. Bangkok (Thailand), December 11–14, 2007, pp. 1–4. https://doi.org/10.1109/APMC.2007.4554877.
34. Masaki, T., Y. Ishii, N. Michishita, et al. Monostatic and Bistatic RCS Measurements for Thin Metasurfaces. In: Proc. of the 2017 IEEE Conference on Antenna Measurements

& Applications (CAMA). Tsukuba (Japan), December 04–06, 2017, pp. 351–352. https://doi.org/10.1109/CAMA.2017.8273448.
35. Walenczykowska, M., W. Bużantowicz, and A. Kawalec. Impact of 3D Model Simplifications on the Determination of Numerical Accuracy of the Radar Cross-Section Values. In: Proc. of the 2024 25th International Microwave and Radar Conference (MIKON). Wrocław (Poland), July 01–04, 2024, pp. 254–259. https://doi.org/10.23919/MIKON60251.2024.10633938.
36. Bużantowicz, W., and M. Walenczykowska. 2025. Impact of 3D model simplifications on the determination of numerical accuracy of the radar cross-section in aerial target recognition issues. *IET Radar, Sonar & Navigation* 19(1):e70003.
37. Kolundžija, B.M., and A. Djordjević. 2022. *Electromagnetic modeling of composite metallic and dielectric structures*. London: Artech House.
38. Pavlović, M.S., M.S. Tasić, B.L. Mrdaković, et al. WIPL-D: Monostatic RCS Analysis of Fighter Aircrafts. In: Proc. of the 2016 10th European Conference on Antennas and Propagation (EuCAP). Davos (Switzerland), April 10–15, 2016, pp. 1–4. https://doi.org/10.1109/EuCAP.2016.7481481.

9 Simulation-Driven Experimental Testing of Electromagnetic Shielding Properties of Materials

Lukasz Nowicki
Warsaw University of Technology, Warsaw, Poland

Marzena Olszewska-Płacha

Janusz Rudnicki

Małgorzata Celuch
QWED Sp. z o.o., Warsaw, Poland

CONTENTS

9.1 Introduction 144
9.2 Measurement Methods 145
 9.2.1 Single-Post Dielectric Resonator (SiPDR) 145
 9.2.2 Inverted Single-Post Dielectric Resonator—2D Scanner 146
 9.2.3 Dual-Mode Ruby Dielectric Resonator (RuDR) 147
 9.2.4 Q-Choked Split Cylinder Resonator (Q-SCR) 148
 9.2.5 New Q-Choked Sapphire Sandwiched Resonator (Q-SSR) 149
 9.2.6 Split-Post Dielectric Resonator (SPDR) 151
9.3 Sample under Tests 151
 9.3.1 Carbon Coatings on Quartz 152
 9.3.2 Graphene-Based Polymer Composites 153
 9.3.3 Graphene Oxide and Reduced Graphene Oxide 155
 9.3.4 Copper Foils—Conductivity versus Surface Roughness 156
 9.3.5 Thick Sapphires Measurement with Q-SCR 157
 9.3.6 Polylactic Acid for Testing of Multiresonance Feature—Q-SSR 157
9.4 Conclusions 158

DOI: 10.1201/9781003739234-11

9.1 INTRODUCTION

Electromagnetic (EM) warfare and modern defense systems rely heavily on the ability to both detect and evade detection. Surveillance radars, target acquisition systems, and missile guidance platforms depend on precise knowledge of the EM properties of objects they are intended to identify. Similarly, stealth strategies aim to exploit materials engineered to minimize radar cross-section (RCS) through absorption or redirection of EM waves.

Accurate characterization of such materials is essential not only during development but also for quality assurance in industrial applications. However, conventional characterization methods, typically involving large quasi-free-space measurement setups, are both resource- and time-intensive. As the demand for novel radar-absorbing materials grows, so does the need for compact, efficient, and versatile characterization techniques.

The quasi-free-space approach has long been the standard for EM material characterization. In this methodology, material panels are illuminated by broadband or narrow-band antenna systems in environments that mimic their operational use—such as anechoic chambers or open-range test sites. While these setups are effective for real-world validation, they are far from ideal for routine or iterative material development due to

1. *Size and Cost*—Anechoic chambers require significant space and investment in infrastructure;
2. *Complex Calibration*—Ensuring reliable data requires precise calibration, environmental control, and often repeated measurements;
3. *Limited Accessibility*—Few institutions can afford the facilities necessary for full-scale radar signature testing.

These limitations hinder innovation in the development of advanced shielding or absorbing materials, especially in early-stage research and development settings.

To overcome the above constraints, the team at QWED has developed an alternative method for EM characterization using resonant structures. Unlike radiating systems, these rely on the confinement of EM energy within a resonator cavity. When a material sample is inserted into the resonator, it perturbs the established field distribution, leading to measurable changes in the resonant frequency and the quality factor (Q-factor) of the system.

The key advantages of this method include:

1. *Compact Form Factor*—Resonators can be desktop-sized and do not require large anechoic environments;
2. *High Sensitivity*—Resonance-based setups can detect small perturbations, enabling precise measurement of material parameters;
3. *Simulation-Aided Design*—Resonators are designed using full-wave EM simulations, ensuring high accuracy and reproducibility.

9.2 MEASUREMENT METHODS

Resonant techniques utilize the core concept that introducing a material into a resonator modifies its EM field distribution, which, in turn, shifts its resonant frequency and Q-factor. These shifts are closely tied to the material's dielectric constant and electrical conductivity. The strength of resonant methods lies in their precision—they can sensitively detect even minor changes, making them ideal for characterizing thin films or lossy materials. For all resonant-based approaches discussed, data interpretation relies on retro-modeling—a simulation-driven method that fits experimental results to theoretical models. This process involves simulating the resonator in both unloaded and sample-loaded conditions using fast numerical solvers [1], such as the conformal Body-of-Revolution Finite-Difference Time-Domain (FDTD) algorithm [2]. Resonant frequencies and Q-factors are extracted from measured data using advanced signal processing techniques like the Prony method [3], ensuring accurate retrieval of the material's EM properties.

9.2.1 SINGLE-POST DIELECTRIC RESONATOR (SiPDR)

The SiPDR [4] operates at a nominal frequency of 5 GHz and is particularly effective for analyzing samples where the imaginary part (loss factor) dominates over the real part of permittivity. It features a dielectric post enclosed within a metallic cavity. When a sample is introduced, its interaction with the resonator's field produces measurable changes in the resonance behavior—specifically, in frequency and Q-factor. These variations are analyzed to derive properties such as surface resistivity or conductivity. This method is widely used in the semiconductor field for characterizing both intrinsic and doped semi-insulating materials. It is also suitable for studying thin films and composites containing conductive particles like graphene nanoplatelets (GNPs), carbon nanotubes, or metallic powders such as silver (Fig. 9.1).

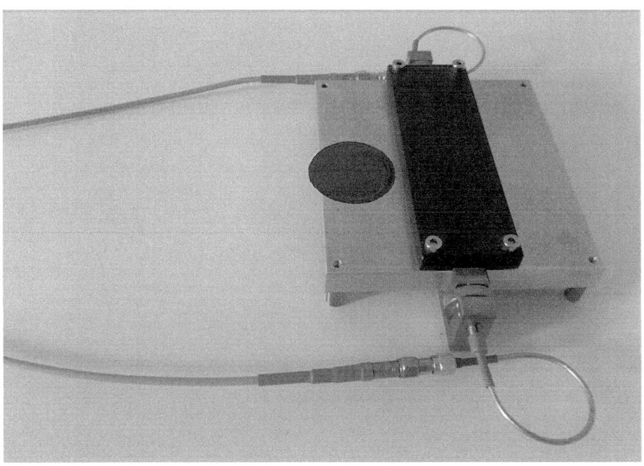

Figure 9.1 Graphene paper sample under test on 5-GHz SiPDR measurement setup which also included a vector network analyzer (VNA) and a laptop with dedicated software.

9.2.2 INVERTED SINGLE-POST DIELECTRIC RESONATOR—2D SCANNER

Electrical characterization is performed using a 2D inverted Single-Post Dielectric Resonator (iSiPDR) [5], a variation of the conventional SiPDR originally developed for the evaluation of semiconductor wafers [6]. The inverted architecture, developed within the framework of the H2020 NanoBat project [7], allows for integration with a 2D scanning stage, enabling the generation of spatial maps of material properties over areas up to 10 × 10 cm. These spatial maps are essential for assessing material homogeneity, detecting fabrication-related defects, and improving material performance. The technique enables non-destructive, contactless assessment of electrical properties and is particularly well-suited for conductive and low-resistivity samples. The iSiPDR operates based on the measurement of resonant frequency shifts and Q-factor variations induced by the presence of the sample, with operation nominally centered at 10 GHz. The measurement procedure comprises two primary stages. Initially, a reference measurement is conducted with the resonator unloaded to establish a baseline response. Subsequently, the sample is placed on the scanning stage, and point-by-point measurements are conducted using a vector network analyzer (VNA) or Q-Meter [8], controlled by dedicated software. Resonant responses are acquired at multiple surface locations, providing a set of frequency and Q-factor data for further analysis.

The measurement system, illustrated in Fig. 9.2, includes the 2D iSiPDR scanner, a Q-meter, and a laptop equipped with software for scanner control and parameter extraction. The Q-Meter records complex scattering parameters (S-parameters), which

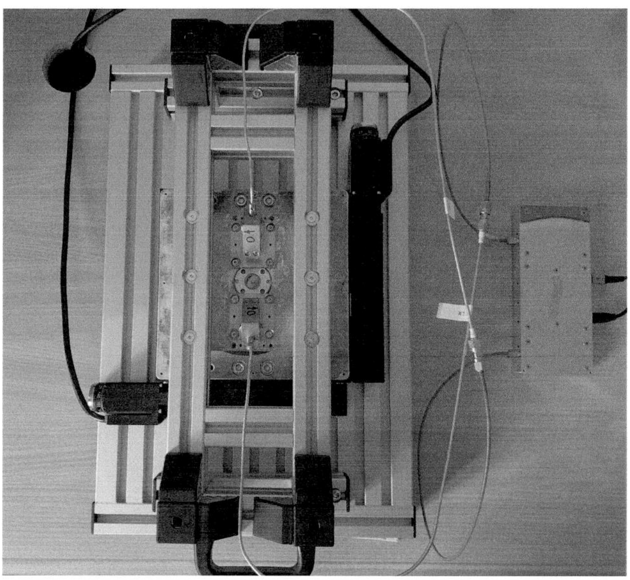

Figure 9.2 A measuring station consisting of 10-GHz iSiPDR scanner, Q-Meter, and laptop with dedicated software.

are used to analyze the resonator's response. Accurate measurement results require the use of flat and uniform samples.

9.2.3 DUAL-MODE RUBY DIELECTRIC RESONATOR (RuDR)

The measurement system uses a cylindrical RuDR [9] connected to a VNA, specifically the Keysight Streamline P5008B model. The VNA is responsible for acquiring scattering parameters, which reflect the resonator's behavior under varying conditions. Signal transmission between the resonator and the VNA is achieved using coaxial cables, ensuring stable, low-loss connectivity. The VNA communicates with a computer via USB interface, enabling real-time data acquisition, analysis, and visualization. An overview of the complete measurement setup is presented in Fig. 9.3.

While the measurement duration may vary slightly based on the VNA configuration, a full acquisition cycle—including sample placement and software configuration—typically requires less than one minute, demonstrating the system's operational efficiency. The RuDR has been developed specifically for the precise characterization of electrically thick conductive layers, such as copper foils, by prioritizing the reduction of dielectric losses. In conventional resonator designs, metallic losses are typically the focus, with dielectric losses considered secondary. In contrast, the RuDR employs a ruby-based resonator structure to minimize dielectric contributions, thereby enabling accurate evaluation of the conductive layers. The system operates in a dual-mode configuration, utilizing TE_{011} and TE_{021} modes, corresponding to approximate resonant frequencies of 13.8 GHz and 20.4 GHz, respectively. This dual-mode approach enhances measurement reliability and extends applicability across a broader frequency range.

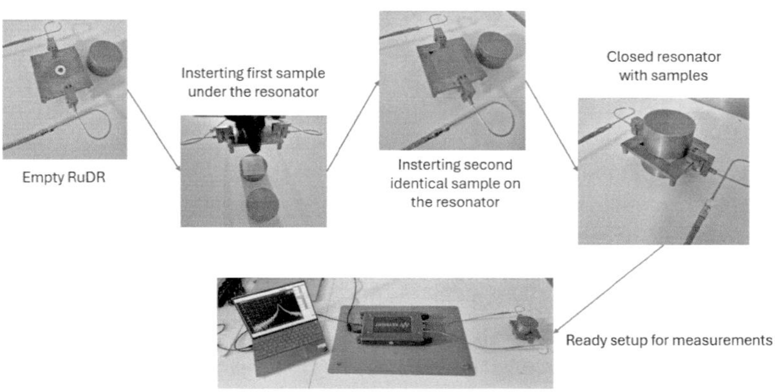

Figure 9.3 The measurement procedure involves three main steps: preparing a sample to fill the resonance cavity, placing samples above and below the cavity (indicated by red arrows) and pressing them with brass cylinders, and connecting the setup via radio frequency (RF) coaxial cables to a vector network analyzer (VNA) to extract material parameters using automated software.

At the core of the resonator is a ruby dielectric element housed within a metallic cavity. The cavity features interchangeable top and bottom walls, which are replaced by the conductive samples being tested. Owing to ruby's ultra-low dielectric loss, internal losses within the resonator are minimized, ensuring that the measured losses predominantly reflect the properties of the conductive materials. The measurement configuration is optimized for electrically thick samples, adhering to the condition $h_s > 3\delta$, where h_s represents the sample thickness and δ denotes the EM skin depth. Compliance with this condition allows the RuDR to target bulk electrical characteristics rather than surface-level effects. Two identical specimens of the material under test, each with a minimum lateral dimension of 23×23 mm, are placed symmetrically on the dielectric element. Uniform contact is ensured by applying pressure using calibrated weights, which helps reduce errors associated with misalignment or nonuniform placement. The VNA records the resonant frequencies and Q-factors for each operational mode. The resonant frequency provides information on EM field distribution within the resonator, while the Q-factor quantifies the balance between stored and dissipated energy, reflecting overall energy efficiency.

The dual-mode functionality not only expands the frequency coverage of the measurements but also allows cross-validation between the two modes, increasing the robustness of the results. This methodology ensures minimal dielectric influence and precise isolation of metallic losses, making the RuDR a highly effective instrument for the EM characterization of conductive materials in high-frequency applications.

9.2.4 Q-CHOKED SPLIT CYLINDER RESONATOR (Q-SCR)

A series of Q-SCRs [10] is under development for operation across the 10–50 GHz frequency range. The initial prototype targets 10 GHz, selected due to its lower manufacturing cost, which facilitates early-stage concept validation. Additionally, this frequency aligns with reference measurements from round-robin testing campaigns [11] and enables direct comparison with existing 10 GHz SPDR data.

The 10 GHz Q-SCR is constructed using copper with a conductivity of $\sigma = 5 \times 10^7$ S/m for the cavity walls that are exposed to the EM field. To introduce controlled losses, choke pockets are fabricated from conductive PLA (Capifil Conductive 750g). Excitation is provided via inductive coupling loops and signal transmission is analyzed across the test fixture. A complete 3D model of the empty Q-SCR (without sample) is constructed to validate the Q-Choke concept (Fig. 9.4). The simulation mesh employs a basic resolution of 0.4 mm, refined to 0.1 mm near the midplane where samples are to be introduced, yielding a total of 921,188 FDTD cells. Simulation durations to obtain the transmission curves presented in Fig. 9.4 are approximately 547 seconds. Figure 9.4 compares the simulated transmission responses of a conventional SCR (without a choke) and the proposed Q-SCR (with a Q-Choke). In the considered frequency range (20% of the center frequency), the only mode within 10 dB of the measurement mode is TE_{012}. It will not perturb or confuse measurements, as it will remain above the TE_{011} resonance. Effectiveness of the Q-Choke in suppressing undesired modes is thereby computationally demonstrated, which makes

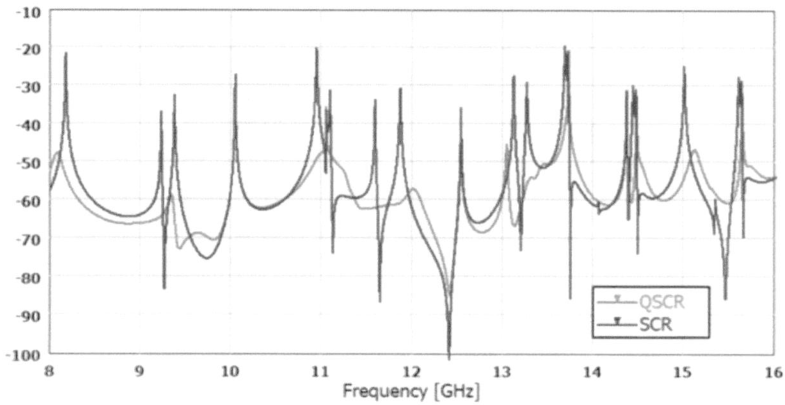

Figure 9.4 Simulated transmission spectrum of classical SCR and new Q-SCR, depicting the excited modes and confirming the suppression of undesired modes by the Q-Choke.

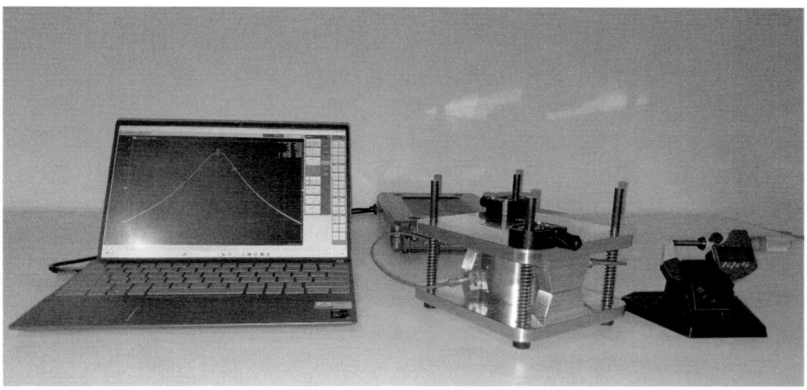

Figure 9.5 Measurement setup featuring the 10-GHz Q-SCR prototype (on the right).

Q-SCR a promising candidate for measuring thick and also high-permittivity material samples (Fig. 9.5).

9.2.5 NEW Q-CHOKED SAPPHIRE SANDWICHED RESONATOR (Q-SSR)

The Q-SSR [12] configuration (illustrated in Fig. 9.6) constitutes a recent development in cylindrical resonator-based material characterization, integrating features from both the established SPDR and the more recent Q-SCR. All three resonator types share a cylindrical geometry and typically operate in TE_{0np} modes, wherein the electric field is oriented in the azimuthal direction within the resonator plane. In traditional SCR designs, such as the Q-SCR, the resonant frequency is primarily governed by the cavity's physical dimensions, with the introduction of a sample resulting in frequency shifts due to perturbations in the EM field.

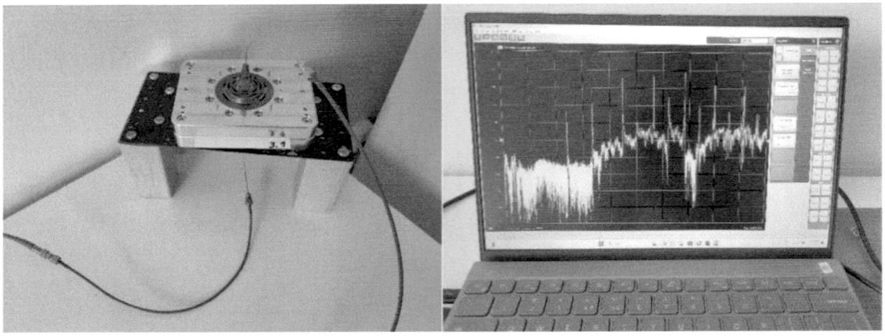

a) b)

Figure 9.6 Q-SSR fixture: (a) with sample inserted into air slot and (b) transmission parameter |S21| shown on vector network analyzer's (VNA) application.

Standard SCR-based sample insertion methods involve mechanically separating the cavity into two halves, placing the sample between them, and reassembling the structure—typically using a lever mechanism. Although this approach ensures direct interaction between the sample and the resonant field, it may lead to measurement variability due to potential misalignment, inconsistencies in sample placement, or uneven contact pressure.

To address these limitations, the Q-SSR introduces structural and functional enhancements aimed at improving field uniformity and measurement repeatability. Unlike SPDR designs, which employ small dielectric posts, the Q-SSR utilizes flat sapphire plates as structural interfaces. Sapphire is selected due to its exceptionally low dielectric loss and high Q-factor, which contribute to improved energy confinement and enhanced measurement accuracy. In this configuration, the test sample is positioned between two sapphire plates, forming a resonant structure referred to as the Sapphire Sandwiched Resonator (SSR). This arrangement promotes a more uniform electric field distribution and reduces uncertainties, particularly when characterizing thin dielectric samples.

A distinguishing feature of the Q-SSR, relative to the Q-SCR, is its advanced Q-Choke design. The designation "Q" signifies the incorporation of a choke mechanism intended to suppress undesired resonant modes and enhance Q-factor stability. In the Q-SCR, the Q-Choke consists of a single internal corrugation forming an air channel that terminates in a lossy material embedded within the cavity sidewalls. While effective, this basic choke structure has been significantly enhanced in the Q-SSR. The Q-SSR employs a more complex multi-ring choke system comprising five concentric corrugations: four embedded within the upper and lower cavity covers, and one additional external ring designed to further limit energy leakage. This advanced configuration offers improved mode suppression and greater resonance stability, particularly beneficial in reducing the influence of environmental disturbances. By optimizing the resonator's internal EM field structure, the Q-SSR enables more accurate

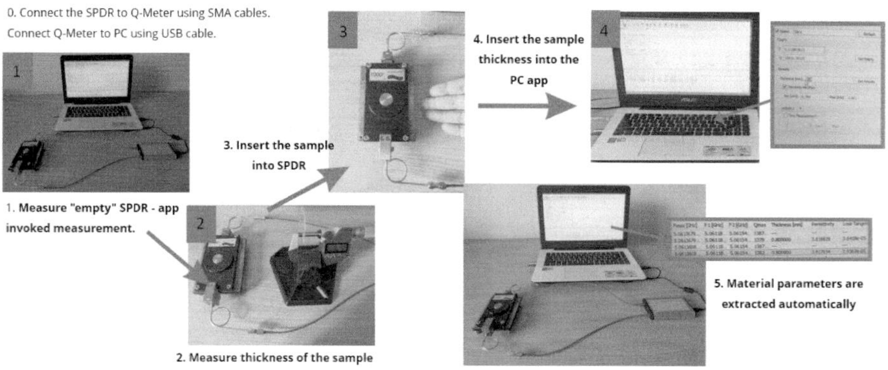

Figure 9.7 5-GHz Split-Post Dielectric Resonator (SPDR) measurement procedure.

and reliable measurements—especially for thin dielectric materials that are sensitive to slight variations in field distribution.

9.2.6 SPLIT-POST DIELECTRIC RESONATOR (SPDR)

The associated measurement procedures are generally straightforward, particularly in open resonator configurations where the sample can be inserted without requiring disassembly of the system. Among these, the SPDR [13] (Fig. 9.7) offers a particularly efficient and minimally complex approach. This configuration enables the measurement of permittivity and loss tangent. Furthermore, spatial mapping of these parameters across the sample surface is possible through the use of a scanning system. While the scanning procedure involves slight additional complexity—such as motorized positioning of the measurement head—user input is limited to simple actions in the control software, such as entering coordinates or selecting predefined options.

The SPDR design consists of a dielectric post situated between two conductive plates, with an air gap or holder to accommodate the sample under test (SUT). Inserting the SUT alters the EM field distribution within the resonator, resulting in a shift of the resonant frequency f_0 and modification of the Q-factor. These changes are directly related to the material's complex permittivity and loss tangent.

9.3 SAMPLE UNDER TESTS

The techniques described here have direct implications for the development and deployment of materials in defense systems. Examples include:

1. *Radar Absorbing Materials (RAMs)*—Characterization of lossy coatings used on aircraft or drones;
2. *EM Shielding Materials*—Evaluation of conductive fabrics, foams, and composites for use in communication shelters;

3. *Stealth Technologies*—Measurement of multilayer composites designed to attenuate radar signals;
4. *Civilian Applications*—Testing of materials used in automotive radar systems, medical devices, or wireless communication shielding.

Examples of practical use of the above methods are described below.

9.3.1 CARBON COATINGS ON QUARTZ

These samples were developed by MateriaNova [14] and IONICS [15], our partners within the M ERA.NET I4Bags project.

Carbon thin films were deposited on a 3-inch quartz wafer using High-Power Impulse Magnetron Sputtering. All the deposition experiments were performed at room temperature with the substrate holder at the floating potential using pure Ar gas (99.99% purity). The film deposition has been performed in dynamic mode using a linear conveyor to pass the substrates to either side of the plasma zone. The number of passes and the conveyor speed have been to deposit a uniform thickness onto the quartz substrates. The target was a rectangular 12 $\times$ 3 inch pure metallic C target. The cathode was located at the top of the chamber and was supplied by a pulsed current source (Starfire impulse power module). A maximum target voltage of –725 V has been applied, delivering a peak current of 150 A. The pulse duration has been set to 30 μs and the pulse frequency adjusted to 600 Hz, giving an average power of 500 W.

Before all the experiments the chamber was pumped down to 1E-07 Torr using 2 turbomolecular pumps. During the deposition, the working pressure was fixed to 5 mTorr. The quartz substrates were cleaned using RBS 25 and then rinsed with ultrapure water. The substrate was not intentionally heated, and the distance between the target and the substrate was fixed at 60 mm.

Four samples were prepared for which the thickness of carbon films was kept constant (200 nm) by controlling the deposition time according to predetermined deposition rates calibrated with a Dektak XT mechanical profilometer.

Nitrogen implantation was performed using a micro-implanter [15]. The process is based on the use of an Electron Cyclotron Resonance (ECR) ion source to produce a highly ionized plasma where nitrogen gas is totally dissociated into multicharged ions (N+, N2+, N3+, ...). The gas injection in the plasma chamber is controlled by a very accurate microvalve to allow precise control of the ion current density arriving on the samples. The plasma working pressure is 1E-06 Torr. The ions are extracted from a plasma chamber by applying a positive voltage up to 40 keV and accelerated toward the samples to be treated. The kinetic energy and therefore the penetration depth of the ions are directly proportional to their charge.

Physical Vapor Deposition (PVD) carbon coatings were implanted with nitrogen at different fluence and for different extraction voltages between 20 and 35 keV. The results obtained on iSiPDR Scanner are shown in Fig. 9.8. The ABS/GNP samples are shown in Fig. 9.9, while selected results are given in Fig. 9.10.

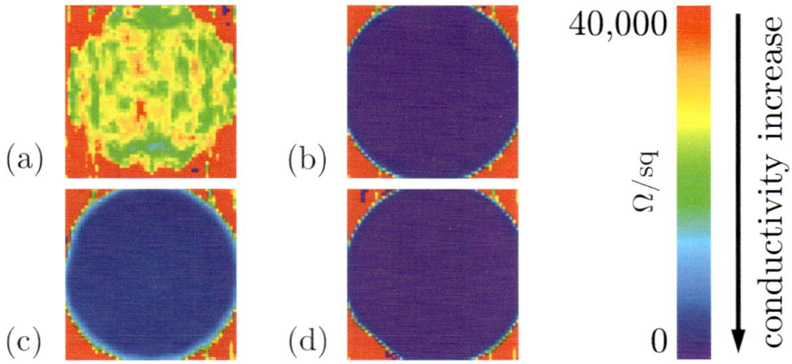

Figure 9.8 Surface resistance maps obtained for: (a) reference sample without ion implantation; (b) ion-implanted sample with nitrogen dose equal to 1E+17 ions/cm^2 of 35 keV; (c) with the same nitrogen dose with 20 keV; and (d) with nitrogen dose equal to 2E+17 ions/cm^2 and 35 keV.

Figure 9.9 Acrylonitrile butadiene styrene with graphene nanoplatelet (ABS/GNP) with different mixture type and GNP concentration.

9.3.2 GRAPHENE-BASED POLYMER COMPOSITES

A series of samples were prepared by our partners at the Warsaw University of Technology [16] using commercially available acrylonitrile butadiene styrene (ABS) and GNPs characterized by an average lateral dimension of 25 μm and a surface area in the range of 120–150 m^2/g. The materials were mixed at a filler content of 10 wt% using various mixing techniques, including the Turbula T2F mixer. A comparative study was conducted to assess the influence of the mixing method and the proportion

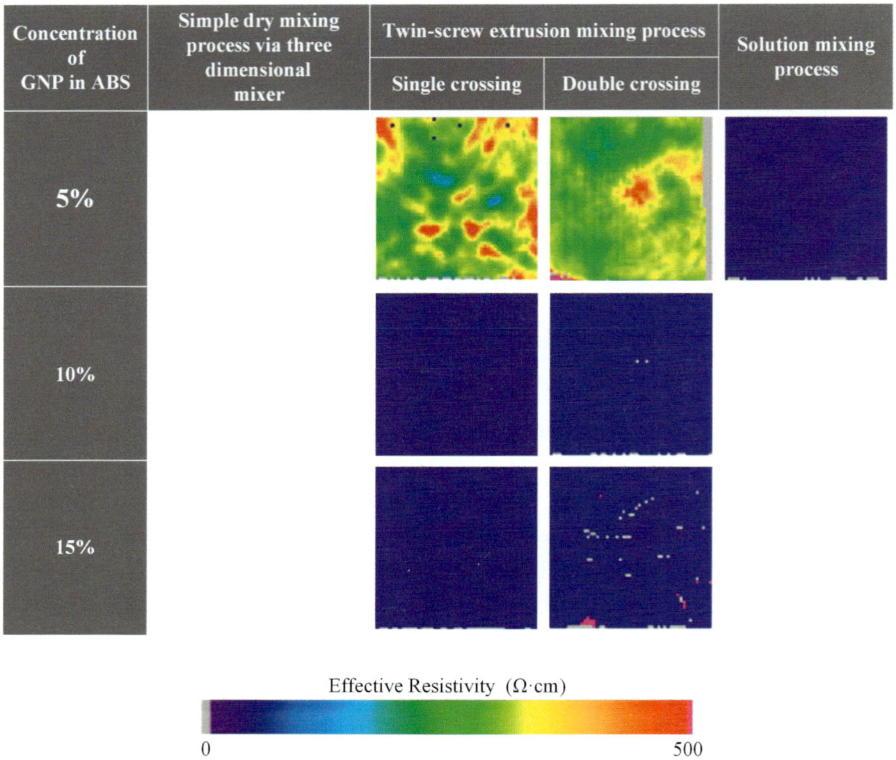

Figure 9.10 Maps of effective resistivity got on 2D iSiPDR for Scanner acrylonitrile butadiene styrene with graphene nanoplatelet (ABS/GNP) with different mixture type and GNP concentration.

of carbonaceous content in the solvent on the final material properties.

Following the mixing process, the resulting powder mixtures were subjected to hot pressing under controlled pressure and temperature conditions using a hydraulic press, leading to the formation of nanocomposite samples suitable for further evaluation.

Non-destructive scanning was performed using a resonator-based technique operating at 10 GHz. The measurements revealed notable spatial variations in extracted resistivity values across the surface of the samples, which indicate substantial non-uniformities. These irregularities were attributed both to the inherent inhomogeneity in the dispersion of the graphene-based fillers and to variations in the thickness of the deposited layers. Mechanical deformations occurring over time, due to internal stresses within the material, caused certain regions to lift away from the substrate. In areas where the material was thicker, the increased interaction volume with the resonator led to a greater detuning effect, which was interpreted as higher dielectric losses.

9.3.3 GRAPHENE OXIDE AND REDUCED GRAPHENE OXIDE

Accurate characterization of graphene-based materials, including graphene oxide (GO) and reduced graphene oxide (rGO), requires the careful preparation of consistent and high-quality samples. The procedures described below detail the synthesis and processing methods applied to obtain reproducible GO and rGO materials for subsequent measurement and analysis, as developed and applied by the Institute of Microelectronics and Photonics, Lukasiewicz Network, within the M ERA.NET I4Bags project.

GO was synthesized via a modified Hummers method, a widely utilized approach involving the oxidative treatment of graphite. This method produces GO sheets enriched with oxygen-containing functional groups, which enhance their dispersibility in water and polar solvents. The synthesis commenced with the oxidation of graphite powder in concentrated sulfuric acid, which served as the reaction medium. The mixture of graphite and H_2SO_4 was maintained at 4 °C to moderate the reaction rate and ensure safe reagent handling. Potassium permanganate was incrementally added to avoid excessive exothermic response, and the mixture was stirred for 30 minutes to promote homogenization. Subsequently, the reaction temperature was elevated to 40 °C and held for 30 minutes to advance the oxidation. Demineralized water was then introduced, followed by an increase in temperature to 90 °C for 15 minutes to support further oxidation reactions. Upon completion, the reaction was terminated by cooling the mixture to ambient temperature and diluting with water. The resulting GO was collected via centrifugation and subjected to multiple dialysis cycles to eliminate residual ions and reaction by-products, yielding high-purity material suitable for advanced characterization.

The purified GO was redispersed in Milli-Q water to form a stable suspension. A vacuum filtration technique was employed for film formation, using a filtration apparatus equipped with a hydrophilic membrane. Equal pressure was applied across the membrane to ensure uniform deposition of the GO sheets. After filtration, the dried GO films were detached from the membrane and prepared in two mass densities: 1.96 mg/cm^2 and 2.21 mg/cm^2. To obtain reduced GO, the GO films underwent chemical reduction using ascorbic acid (Vitamin C) as the reducing agent. The GO samples were immersed in an aqueous solution of ascorbic acid and subjected to controlled heating conditions to facilitate the removal of oxygen-containing functional groups. Following reduction, the rGO samples were thoroughly rinsed with deionized water and dried, resulting in conductive films with enhanced electrical properties.

The preparation of GO and rGO samples involves precise control over synthesis and deposition processes to ensure reproducibility and quality. By varying the mass density and employing chemical reduction, we obtained a set of samples suitable for detailed EM characterization. These samples serve as the foundation for exploring the dielectric and conductive properties of graphene-based materials, which will be examined in the subsequent sections.

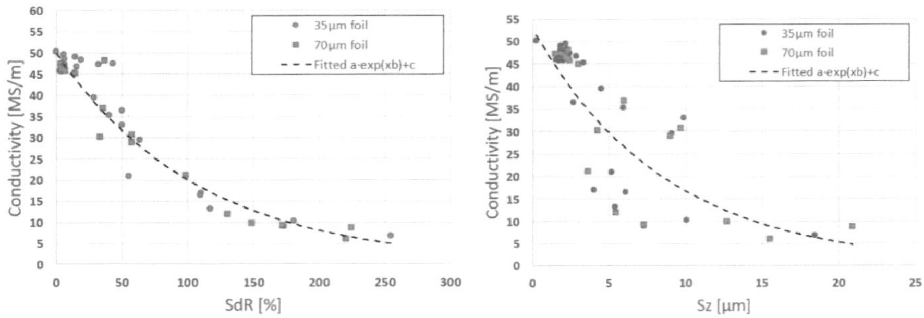

a) b)

Figure 9.11 Correlation between effective conductivity σ (measured at 21 GHz in the RuDR test fixture and surface roughness parameters: (a) SdR—developed interfacial area ratio and (b) Sz—maximum height.

9.3.4 COPPER FOILS – CONDUCTIVITY VERSUS SURFACE ROUGHNESS

The copper foils analyzed in this study were produced using standard industrial manufacturing processes at Circuit Foil Luxembourg [17]. The foils were delivered in two thicknesses: 35 μm and 70 μm. Each foil is identified by its thickness and various roughness parameters measured using a non-contact laser interferometry [18] (providing Sz, Sa, and Sdr values). The initial set of samples included 24 types of foil, with three sheets of each type, representing different combinations of foil base, applied treatment, and thickness (35 μm or 70 μm). Each sheet was provided in A4 paper format, from which 60 $\times$ 60 mm samples were cut, yielding a total of 72 copper foil samples for analysis.

In the first phase of the study, foil loss was measured using a RuDR at 21 GHz. For each measurement, two nominally identical samples, cut from the same foil sheet, were required and mounted as illustrated in Fig. 9.2 to form the two bases of the cylindrical ruby resonator. The samples were securely pressed by metallic blocks. The RuDR test fixture was connected to a VNA, which determined the Q-factor from the 3 dB bandwidth and resonance frequency. A dedicated software application then converted this data into effective conductivity σ values using full-wave modeling. Each of foil was measured 16 times via application function which do single measurement every 5 seconds. For further consideration we will use an average values of conductivities.

The effective conductivity σ was measured on the electrolyte side of each foil for both thicknesses (35 μm and 70 μm). These conductivity σ values were then correlated with surface roughness R_s parameters, showing a decreasing exponential relationship in all cases, as illustrated in Fig. 9.11. To better visualize this trend, an exponential curve was fitted to the entire set of measurements of the RuDR, which suggests that the foil thickness does not influence conductivity σ. Although a few data points deviate from the fitted curve, it is important to note that conductivity σ is not only affected by surface roughness R_S; other factors, such as the grain size of the copper foils, also play a role. The impact of grain size is currently being investigated as part of the 5GFoil project.

9.3.5 THICK SAPPHIRES MEASUREMENT WITH Q-SCR

Capability of making unambiguous measurements of samples thicker than allowed in the pre-existing techniques is first validated. A batch of five sapphire samples is considered, each of ca. 0.4 mm thickness. The expected value of relative in plane permittivity is 9.3954, based on the reference measurements with whispering gallery modes for ultra-low-loss dielectric materials [19]. According to the brochures [20] and open literature, even a single such sample exceeds the maximum thickness limitations of commercially available SCR [21] and Fabry-Perot Open Resonator (FPOR) [22] instruments. In a 10-GHz SPDR a single sample can be measured, giving $\varepsilon_S = 9.390$, $\tan(\delta_S) = 1.21 \times 10^{-5}$. In the new 10-GHz Q-SCR, all five samples are stacked and successfully measured. Averaged value of ε for 1÷5 disks (5 measurements) is 9.4032 ± 0.0845%.

Very good accuracy of the proposed Q-SCR is also demonstrated for dielectric loss extraction. Measured loss tangent values are $\tan(\delta_S) = 1.37 \times 10^{-5}$, 1.89×10^{-5}, 1.83×10^{-5}, 1.93×10^{-5}, 1.9×10^{-5} for stacks of 1÷5 disks, respectively, showing a spread of $0.16 \div 0.72 \times 10^{-5}$ compared to the value measured in 10-GHz SPDR, which is noticeably below the declared SPDR accuracy.

9.3.6 POLYLACTIC ACID FOR TESTING OF MULTIRESONANCE FEATURE—Q-SSR

A digital twin of the Q-SSR resonator was prepared using EM modeling and simulation software. Given the known geometry and material parameters of the Q-SSR resonator, an ultra-fast conformal BoR FDTD algorithm was employed to model both empty and loaded configurations in a grid-search regime. The advanced Prony method was applied with automatic selection of signal post-processing parameters to extract the fields of the simulated microwave signals.

Numerous simulations were conducted by varying the permittivity of the sample, which provided insights into the changes in resonance frequency. Specifically, an increase in the sample's permittivity resulted in a decrease in the resonance frequency. In the measurement approach, an empty Q-SSR resonator served as a reference, followed by the insertion of a sample with unknown material parameters. By analyzing shifts in resonance frequency and changes in Q-factors, we compared the experimental results with simulated data from the EM simulator. To extract the loss tangent, a variation of sample permittivity was used in conjunction with the calculation of the filling factor [23].

Understanding the influence of the sample on the EM field distribution within the resonator cavity is crucial. A high-permittivity sample accumulates the EM field within itself, causing the field to recede from the copper walls. This leads to an increase in the Q-factor when intrinsic sample losses are excluded, as less field

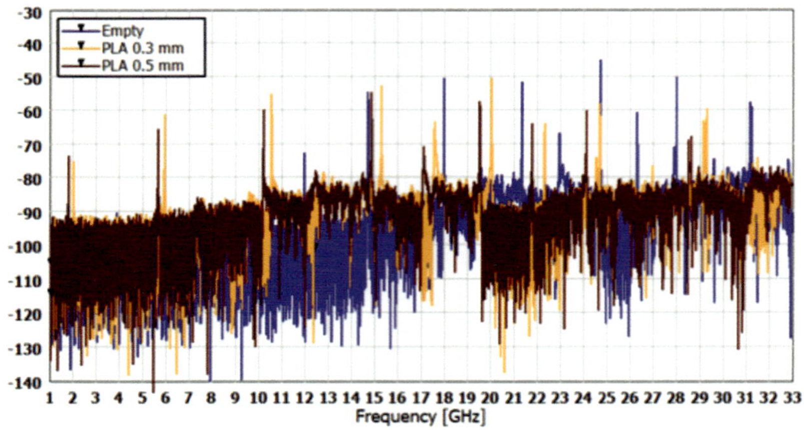

Figure 9.12 Transmission of Q-SSR with inserted PLA of various thickness measured on VNA.

interacts with the lossy copper walls. However, real samples introduce additional losses, which can be extracted and used to calculate the loss tangent $\tan\delta$ once all Q-factor components are separated and known. The Q-SSR resonator features multiple resonance frequencies, making the process more time-efficient compared to single-resonance approaches. However, this also increases the computational effort, necessitating the use of the BoR method. To achieve high resolution and accuracy, each simulation point—defined by a specific permittivity and sample thickness—required approximately two minutes to compute. Initial simulations were conducted over a broad range (11–33 GHz) to identify resonance frequencies, followed by more refined simulations around these resonances to enhance accuracy in frequency and Q-factor determination. In Fig. 9.12 is shown a measurement taken on Q-SSR fixture for samples under test and without any sample inserted into resonator acting as a reference to frequencies and Q-Factors. From all resonances an average value for PLA 0.3 mm is $\varepsilon = 2.60 \pm 0.17\%$ and is $\varepsilon = 2.59 \pm 0.19\%$ for PLA 0.5 mm.

9.4 CONCLUSIONS

Resonant measurement techniques represent a powerful alternative to traditional free-space characterization methods for EM shielding and absorbing materials. By leveraging compact resonator setups, including SPDR, SiPDR, iSiPDR, Q-SCR, and Q-SSR, researchers and engineers gain access to precise, repeatable, and scalable measurements of critical material parameters.

The QWED team continues to expand the capabilities of these techniques, integrating scanning stages, broadband coverage, and simulation-based modeling to support the next generation of stealth and EM-compatible materials.

AUTHOR CONTRIBUTIONS

Conceptualization, M.C. and M.O.; methodology, M.O. and M.C.; software, J.R. and M.C.; validation, M.O.; formal analysis, M.O.; investigation, L.N.; resources, M.C.; data curation, L.N. and J.R.; writing—original draft preparation, L.N.; writing—review and editing, M.C.; visualization, L.N. and J.R.; supervision, M.C.; project administration, M.C. and M.O.; funding acquisition, M.C. All authors have read and agreed to the published version of the manuscript.

DATA AVAILABILITY STATEMENT

The data that support the findings of this chapter are available from the corresponding author, L.N., upon reasonable request.

FUNDING

This work received no specific grant from any funding agency in the public, commercial, or not-for-profit sectors.

REFERENCES

1. Celuch, M., W. Gwarek, and A. Wieckowski. Enhanced Resolution Material Imaging with Dielectric Resonators: A New Implicit Space-Domain Technique. In: Proc. of the 2019 IEEE MTT-S International Microwave Symposium (IMS). Boston (MA, USA), June 02–07, 2019, pp. 55–58. https://doi.org/10.1109/MWSYM.2019.8701021.
2. QuickWave BOR software specification. 2025. https://qwed.eu/quickwave/quickwave-3d/ (accessed September 21, 2025).
3. Kozakowski, P., and M. Mrozowski. 2002. Low-order models from FD-TD time samples. *IEEE Microwave and Wireless Components Letters* 12(11):438–440.
4. Krupka, J., and J. Mazierska. 2007. Contactless measurements of resistivity of semiconductor wafers employing single-post and split-post dielectric-resonator techniques. *IEEE Transactions on Instrumentation and Measurement* 56(5):1839–1844.
5. Olszewska-Płacha, M., A. Masouras, A. Wieckowski, et al. Contactless Device for 2D Imaging and Precise Characterisation of Electrical Parameters of Anode Materials for Battery Cells. In: Proc. of 24th International Microwave and Radar Conference (MIKON 2022). Gdańsk (Poland), September 12–14, 2022, pp. 1–3. https://doi.org/10.23919/MIKON54314.2022.9924947.
6. Krupka, J., K. Derzakowski, T. Zychowicz, et al. 2007. Measurements of the surface resistance and conductivity of thin conductive films at frequency about 1 GHz employing dielectric resonator technique. *Journal of the European Ceramic Society* 27(8–9):2823–2826.
7. NanoBat Project. 2025. https://qwed.eu/projects/nano-bat/ (accessed September 21, 2025).
8. Celuch, M., O. Douheret, P. Korpas, et al. Portable Low-Cost Measurement Setup for 2D Imaging of Organic Semiconductors. In: Proc. of the 2020 IEEE MTT-S International Microwave Symposium (IMS). Los Angeles (CA, USA), August 04–06, 2020, pp. 373–376. https://doi.org/10.1109/IMS30576.2020.9224053.

9. Krupka, J., T. Shimada, I. Ueda, et al. 2011. Electromagnetic properties of sapphire, ruby, and irradiated ruby at frequencies of 30-40 GHz. *Applied Physics Letters* 99(10):102901.
10. Celuch, M., M. Olszewska-Płacha, L. Nowicki, et al. 2024. A novel Q-choked resonator for microwave material measurements alleviating sample thickness limitations of existing techniques. *IEEE Microwave and Wireless Technology Letters* 34(6):845–848.
11. International Electronics Manufacturing Initiative. The iNEMI "5G/mmWave Materials Assessment and Characterization" project. https://www.inemi.org/article_content.asp?adminkey=5cc4f4100ebf2ba1f3e6fd6294749139&article=161 (accessed May 11, 2025).
12. Gwarek, W., M. Celuch, and L. Nowicki. A Novel Q-Choked Sapphire Sandwiched Resonator for Wide-Band Measurements of Flat Dielectric Samples. In: Proc. of the 2025 IEEE MTT-S International Microwave Symposium (IMS), Th1D-5.
13. Krupka, J., A.P. Gregory, O.C. Rochard, et al. 2001. Uncertainty of complex permittivity measurements by split-post dielectric resonator technique. *Journal of the European Ceramic Society* 21(15):2673–2676.
14. Materia Nova. 2025. https://www.materianova.be/en/ (accessed September 21, 2025).
15. Ionics Surface Technologies. 2025. https://www.ionics-group.com/en (accessed September 21, 2025).
16. Nowicki, L., K. Filak, M. Celuch, et al. A Fast Modelling-Based Technique for the Characterization of Graphene-Based Polymer Composites. In: Proc. of the 2023 IEEE MTT-S International Conference on Numerical Electromagnetic and Multiphysics Modeling and Optimization (NEMO). Winnipeg (MB, Canada), June 28–30, 2023, pp. 73–76. https://doi.org/10.1109/NEMO56117.2023.10202288.
17. Circuit Foil Luxembourg. 2025. https://www.circuitfoil.com (accessed September 21, 2025).
18. Enoki, M., M. Watanabe, P. Chivavibul, et al. 2000. Non-contact measurement of acoustic emission in materials by laser interferometry. *Science and Technology of Advanced Materials* 1(3):157–165.
19. Krupka, J., K. Derzakowski, A. Abramowicz, et al. 1999. Use of whispering-gallery modes for complex permittivity determinations of ultra-low-loss dielectric materials. *IEEE Transactions on Microwave Theory and Techniques* 47(6):752–759.
20. QWED Measurement Setups. 2025. https://qwed.eu/hardware-2/ (accessed September 21, 2025).
21. Janezic, M.D., and J. Baker-Jarvis. 1999. Full-wave analysis of a split-cylinder resonator for nondestructive permittivity measurements. *IEEE Transactions on Microwave Theory and Techniques* 47(10):2014–2020.
22. Karpisz, T., B. Salski, P. Kopyt, et al. 2019. Measurement of dielectrics from 20 to 50 GHz with a Fabry–Pérot open resonator. *IEEE Transactions on Microwave Theory and Techniques* 67(5):1901–1908.
23. Krupka, J. 2006. Frequency domain complex permittivity measurements at microwave frequencies. *Measurement Science and Technology* 17(6):R55.

10 To Keep the Skies Safe *Final Conclusions and Remarks*

Jan Pietrasieński

Witold Bużantowicz
Faculty of Mechatronics, Armament and Aerospace,
Military University of Technology, Warsaw,
Poland

CONTENTS

10.1 Military Perspective..161
10.2 Scientific Perspective..163

10.1 MILITARY PERSPECTIVE

Integrated Air and Missile Defense (IAMD) constitutes a critical component of national security for the countries of the eastern flank of the North Atlantic Alliance. This issue has gained particular significance in the context of recent developments: the conflict in Ukraine and the direct threat to the security of the eastern flank. Recently, several unprecedented events have occurred, prompting a deeper and more effective reflection on this topic. Notable examples include, among others: the failure of an S-300 SAM system missile in Przewodów, eastern Poland (November 15, 2022); the first-ever combat employment of an medium-range ballistic missile (MRBM)-class missile against the Ukrainian city of Dnipro (November 21, 2024); and the extensive violations of North Atlantic Treaty Organization (NATO) airspace by Russian drones (September 10, 2025).

This monograph represents the result of intensive discussions on the issues of contemporary air and missile defense conducted by representatives of government, the armed forces, the defense industry, media, and academia, among others, during meetings held as part of the annual Regional IAMD Coalitions Conference [1] as well as other events related to air and missile defense issues. The fundamental conclusion derived from these discussions is that air and missile defense hold the highest priority; moreover, air defense on the Eastern Flank of NATO is a complex issue

DOI: 10.1201/9781003739234-12

that necessitates coordination and the enhancement of capabilities among individual member states.

Recently, there has been a significant increase in the challenges faced by the air and missile defense systems of these countries. This is conditioned by changes in the methods of conducting air attacks, which increasingly rely on the mass deployment of mini unmanned platforms and maneuvering, ballistic, and hypersonic missiles. These offensive measures pose fundamental challenges to counter-UAV and missile defense systems, particularly affecting nations of the Eastern Flank. This is due to the fact that the reaction time to these threats is drastically shorter, necessitating different means and methods of countering them. Specifically, this involves the need for kinetic neutralization of hard targets, such as supersonic and ballistic missiles. In contrast, countering drones is hindered by economically unacceptable limitations; small, low-flying drones are difficult to detect with traditional radars and are too inexpensive to be effectively neutralized with costly missile systems.

In the context of the aforementioned threats, discussions emphasize the fundamental importance of IAMD systems, cf. Figs. 10.1 and 10.2. The development of interoperable, network-centric systems that facilitate the integration of radar and optical sensors, missile armaments, communication systems, and command and control assets is indispensable. A critical aspect in this regard pertains to the detection,

Figure 10.1 On September 16, 2025, Poland conducted its first live-fire test of the US-made Patriot air defense missile system acquired under the WISLA program. The training was conducted at the Central Air Force Training Ground in Ustka, using the IBCS (Integrated Battle Command System) command network (photo: © by Michał Wajnchold, FoxTwo.pl).

Figure 10.2 On June 22, 2023, Poland conducts first live-fire test of Common Anti-air Modular Missile (CAMM) missiles acquired as a part of the MALA NAREW program. Two of the air targets simulated the flight of cruise missiles flying at altitudes of two and three kilometers, while the third simulated the flight of an reconnaissance Unmanned Aerial Vehicle (UAV) at an altitude of 500 meters. All three targets were shot down (photo: © by 16th Mechanized Division).

recognition, and classification of aerial targets. An illustrative example is the IBCS (Integrated Battle Command System), which enables multi-domain integration for the United States military and its strategic allies. Its modular and open architecture allows for seamless integration with both legacy and future systems, thus enhancing operational awareness and supporting decision-making processes related to engagement strategies.

10.2 SCIENTIFIC PERSPECTIVE

Due to the mathematical and physical complexity of the processes involved, defense against aerial and space threats relies critically on tools of mathematical modeling of phenomena and computer simulation. These tools are essential for conducting analyses of the current system configurations' effectiveness and the utility of proposed future solutions. An example of advanced modeling application is the determination of the effective radar cross-section (RCS) of aerial targets, which serves as a fundamental parameter for detection and tracking devices of aerial threats.

Computer modeling and simulation methods represent a highly promising avenue of research for the analysis of complex real-time systems characteristic of air defense systems. Software and hardware simulation tools, in addition to their analytical and cognitive capabilities, form the fundamental basis for the operation of advanced training devices and air defense management systems, with their functionality adapted to meet real-time requirements.

Currently, it is standard practice that each new missile-based air defense system is equipped with a simulator or trainer provided by the manufacturer. The ability to utilize a comprehensive simulation environment is not only a training convenience but also an essential tool for the daily operation of personnel operating air and missile defense systems. These systems enable users to verify, through exercises and experiments, the capabilities for threat neutralization under conditions closely resembling actual combat scenarios. In perspective, these systems should possess a transnational and interoperable character, compatible with solutions adopted by other nations, to facilitate joint operations within a NATO alliance framework in the event of a threat.

This requirement is particularly pertinent for countries of NATO's Eastern Flank.

AUTHOR CONTRIBUTIONS

Conceptualization, J.P. and W.B.; writing—original draft preparation, J.P.; writing—review and editing, J.P. and W.B.; supervision, J.P.; project administration, W.B.; funding acquisition, W.B. All authors have read and agreed to the published version of the manuscript.

DATA AVAILABILITY STATEMENT

Data sharing is not applicable to this chapter as no new data were created or analyzed in this study.

FUNDING

This work received no specific grant from any funding agency in the public, commercial, or not-for-profit sectors.

REFERENCE

1. Regional IAMD Coalitions Conference. 2024-2025. Military University of Technology, Warsaw, Poland. https://wml.wat.edu.pl/iamd/ (accessed September 20, 2025).

Index

absorbing materials, 144, 158
aerothermodynamics, 19, 20
AI, 7, 8, 26, 68, 74–76, 79, 82, 83, 96, 98
air defense, 3, 14, 15, 35, 40, 45, 82, 118, 161–164
air defense missile, 162
air defense system, 14, 15, 163, 164
air superiority, 3, 9
anti-aircraft systems, 120
anti-missile systems, 138
artificial intelligence, 7, 16, 42, 48, 74, 108

battlefield, 4–8, 45, 74, 75, 80, 82

C2, 10, 17, 39
camera, 75, 104, 106
capability development, 39
classification, 35, 104, 105, 110, 112–114, 163
combat efficiency, 4
command and control, 13, 14, 16, 36, 38, 39, 162
computational fluid dynamics, 32
critical decisions, 8, 16
CubeSat, 94–96, 98

data fusion, 75
detection, 10, 14, 25, 32, 75, 78, 80, 81, 83–91, 95–97, 104, 108, 114, 144, 162, 163
drone, 13, 14, 26, 68, 74, 75, 77, 81–83, 91, 97, 98, 104, 151, 161, 162

early warning radar, 10
electronic circuits, 8
EM shielding, 158
experimental fluid mechanics, 30

Fast Multipole Method, 119
Finite-Difference Time-Domain, 119
fire control network, 13
full-wave EM simulations, 144

GBAD, 3, 35, 38–42
Geometrical Optics, 119
geometry simplification, 126
gliding bombs, 7, 9

heat transfer, 24
high-integration circuits, 7
high-speed aerodynamics, 19
hypersonic cruise missiles, 85
hypersonic glide vehicles, 85
hypersonic threats, 9, 10, 85

IAMD, 12, 14–17, 34, 39, 40, 161
IAMD systems, 162
IBCS, 10, 13–17, 163
IBCS (Integrated Battle Command System), 162
integrated air and missile defense, 12–14, 16
Integrated Battle Command System, 13, 163
integration, 7, 10, 15–17, 26, 29, 42, 76, 77, 83, 84, 146, 162, 163, 165
interoperability, 14, 16, 17, 77

large-caliber rockets, 7, 9
local alliances, 6

Method of Moments, 119
mobile fire control radars, 10
mono- and multi-static configurations, 10
multi-domain operations, 17
Multi-Level Fast Multipole Method, 119
multi-static radar, 93

NATO, 10, 15, 19, 20, 27, 32, 35, 41, 43, 98, 161, 164
North Atlantic Treaty Organization (NATO), 161

operational concepts, 4–6
operational deployments, 9
operational requirements, 4, 6

P-18PL, 10, 89
passive location system, 10
PCL-PET, 10
peacetime, 4, 5, 11, 82
Physical Optics, 119
physics-based simulation, 31, 67
PPO, 54, 57–60, 62, 63, 67

quasi-anechoic chamber, 120–123, 126, 134, 136, 137

Radar Cross-Section, 118
RCS, 74, 78, 80, 84, 86–90, 97, 114, 118–138, 144, 163
reaction time, 3, 5, 7, 8, 162
recognition, 8, 14, 68, 74, 75, 78, 83, 85, 97, 104–106, 108–113, 119, 138, 163
reinforcement learning methods, 55, 62
research center, 31
resource allocation, 46, 67
retro-modeling, 145

SAMS, 7, 10
self-protection systems, 7
sensors, 7, 10, 13–16, 30, 42, 75–77, 82, 83, 87, 90, 91, 94, 96, 99, 109, 112, 113, 162
SHORAD, 40
signal processing, 75, 77, 90, 97, 98, 108, 145
spiral development process, 6
stealth technology and jamming, 7
surface-to-air missile systems, 7
swarm formation, 104, 105, 108, 111–114

testing, 5, 9, 10, 19, 23, 25, 41, 67, 126, 127, 144, 148
time-varying threat priorities, 67
tracking algorithms, 80, 81
Transmission-Line Matrix, 119

VHF band radar, 87, 90, 91
VKI, 19–32
Von Karman Institute, 18, 19, 21

wartime, 5
WIPL-D, 120, 121, 124–126, 135